Two-Dimensional
Electrophoresis and
Immunological Techniques

Two-Dimensional Electrophoresis and Immunological Techniques

Bonnie S. Dunbar

Baylor College of Medicine
Houston, Texas

Plenum Press • New York and London

Library of Congress Cataloging in Publication Data

Dunbar, Bonnie S.
  Two-dimensional electrophoresis and immunological techniques.

  Includes bibliographies and index.
  1. Electrophoresis, Polyacrylamide gel. 2. Proteins — Analysis. 3. Immunoglobulins
— Analysis. I. Title.
QP519.9.E434D86   1987   574.19′245   86-30393
ISBN 0-306-42439-8

© 1987 Plenum Press, New York
A Division of Plenum Publishing Corporation
233 Spring Street, New York, N.Y. 10013

Printed in the United States of America

Preface

This text is a summary of basic principles and techniques and is dedicated to all those students who have been told by their mentors, "Go forth and do two-dimensional gels and have the results on my desk tomorrow." No attempt has been made in this text to provide exhaustive lists of references related to basic principles or techniques or to list every company or supplier involved in this area of research. Nevertheless, it is hoped that sufficient information is given to help a new investigator or student appreciate the complexities but develop sufficient expertise to carry out these techniques successfully. The discussions are designed to instill in basic science and clinical investigators of all levels of expertise an appreciation of the power of combining a variety of techniques as well as to provide basic insight into the theories, complexities, and problems frequently encountered with electrophoretic and immunochemical methods.

Bonnie S. Dunbar

Houston

Acknowledgments

I wish to thank my students and staff for their patience and support throughout the preparation of this text. I would like to acknowledge my appreciation for my extensive discussions with Dr. David Sammons (University of Arizona) and to Dr. N. L. and Dr. N. G. Anderson and their colleagues (Argonne National Laboratory) for their invaluable advice and suggestions in this area over the years. I thank my research assistant, Ms. Donnie Bundman, for her technical support in carrying out many of the experiments and working out many of the techniques described in this text. I further thank my graduate students (Ms. Grace Maresh and Messrs. Ken Washenik and Eric Schwoebel), technicians (Ms. Claire Lo and Mr. Steve Avery), research associates (Dr. Therese Timmons and Dr. Sheri Skinner), and colleague (Dr. JoAnne Richards) for their many helpful suggestions in the writing of this text.

I wish to thank Ms. Suzanne Saltalamacchia for her expertise in typing and her help in preparing this manuscript, as well as Mr. David Scharff for his artwork and Ms. Debbie Delmore for assistance with a portion of the photography.

I extend my appreciation to my high school science instructor (Mr. James Carson) and to my graduate student advisors (Dr. Paul Winston, Dr. Charles B. Metz, and Dr. Joseph C. Daniels, Jr.) for their instruction during my early years as a research scientist.

Finally, a very special thanks to my parents, who have always instilled in me the importance of learning.

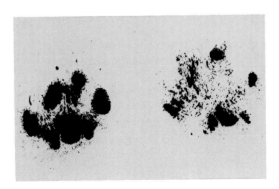

Contents

Chapter 2

Applications and Strategies for Use of Polyacrylamide Electrophoresis

Chapter 3

Sample Preparation for Electrophoresis

Chapter 4

Protein Detection in Polyacrylamide Gel Electrophoresis

Chapter 5

**Basic Principles of Posttranslational Modification of Proteins and
Their Analysis Using High-Resolution Two-Dimensional
Polyacrylamide Gel Electrophoresis**

Chapter 6

Isotopic Labeling of Proteins for Electrophoretic Analysis

Chapter 7

Use of Autoradiography in Polyacrylamide Gel Electrophoresis

Chapter 8

**Strategies for Use of Polyacrylamide Gel Electrophoresis in the
Preparation and Characterization of Antibodies**

Chapter 9

Practical Methods for Laboratory Photography

Chapter 10

Troubleshooting and Artifacts in Two-Dimensional Polyacrylamide Gel Electrophoresis

Chapter 11

Advances in Technology of High-Resolution Two-Dimensional Polyacrylamide Gel Electrophoresis

reaction will take place at each of the platinum electrodes. Usually, hydrogen (H_2) is evolved or some metal is deposited at the electrode (termed the *cathode*), which is connected to the *negative* pole of the battery. During electrolysis, a nonmetal (e.g., O_2) is liberated at the *anode,* which is at the *positive* pole of the battery. The usual reactions that occur in an electrophoresis chamber are as follows:

1. *Cathode reactions* (where reduction or the gain of electrons occurs):

$$2e^- + 2H_2O \rightarrow 2OH^- + H_2$$

$$HA + OH^- \rightarrow A^- + H_2O$$

2. *Anode reactions* (where oxidation or the loss of electrons occurs)

$$H_2O \rightarrow 2H^+ \rightarrow \tfrac{1}{2}O_2 + 2e$$

$$H^+ + A^- \rightarrow HA$$

In order to maintain the electric current, it is necessary to have a complete circuit (a closed-loop system in which the electric charge can return to its starting point). If this complete circuit has an electrolytic conductor, a chemical reaction must occur at the electrodes. An example

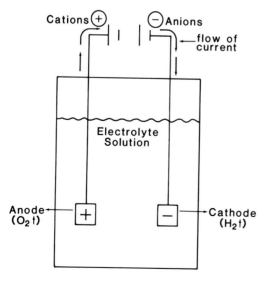

Figure 1.1. Diagram of an electrolytic cell. (Modified from Murphy and Rousseau, 1969.)

of an electrolysis unit (i.e., electrolytic cell) is shown in Figure 1.1.

As illustrated in Figure 1.1, the ions ($+$ and $-$) are free to move in an electrolytic cell. If a battery is attached to this cell, it will generate an electric field that pushes electrons through the wires in the directions given by the arrows. The rate at which electricity moves in a circuit is measured in *amperes* (A). A current of 1 A corresponds to the transfer of electricity at a rate of 1 coulomb/second. The term *coulomb* is the measurement for the quantity of electricity required to deposit 1.118×10^{-3} g of silver from a solution of silver nitrate.

Faraday further determined that the mass of any substance formed by the passage of a given amount of electricity is directly proportional to the equivalent weight of that substance. The liberation of one *equivalent weight* of any element during electrolysis requires 1 Faraday (F), which is 96.493 coulombs. A Faraday is also the charge on a mole of electrons. Because every electron has an identical negative charge, neutralization of each positive or negative charge during electrolysis requires the gain or loss of one electron. The number of electrons (Faradays) that will neutralize the total charge on 1 mole of singly charged atoms will therefore neutralize ½ mole of doubly charged atoms. For example

$$Ag^+ + e^- \rightarrow Ag^\circ, \qquad \text{equivalent weight} = \text{atomic weight}$$

$$Cu^{2+} - 2e^- \rightarrow Cu^\circ, \qquad \text{equivalent weight} = \text{atomic weight}/2$$

$$O_2^- \rightarrow \tfrac{1}{2}O_2 + 2e^-, \qquad \text{equivalent weight} = \text{atomic weight}/2$$

In electrolysis reactions, the equivalent weight of an element is therefore, equal to its atomic weight divided by the number of electrons it gains or loses (from Murphy and Rousseau, 1969).

When an electrolyte is dissolved in water, its molecules dissociate into oppositely charged fragments (*ions*). The conduction of electricity occurs when electrodes connected to a battery are placed in an electrolyte solution and the positive ions (cations) migrate to the negatively charged cathode, where each cation will pick up one or more electrons. Meantime, the negatively charged ions (anions) will lose their negative charge by transferring electrons to the positively charged anode. Electrons are therefore "pumped" out of solution at the anode through the external circuit and back into solution at the cathode; thus the current continues to flow as long as positive and negative ions are present in the solution. The voltage of this reaction is a measure of the electromotive force of this system (Murphy and Rousseau, 1969).

III. MOVEMENT OF MOLECULES IN AN ELECTRICAL FIELD

When any molecule is placed in an electric field, a *force* is exerted on it, which depends on both the *strength* of the electrical field as well as the *charge* of that molecule (see discussions by Cooper, 1977; Freifelder, 1976). The mathematical equation used to express this phenomenon is

$$F = (E/d)\,(q)$$

where F is the force, E is the potential difference between electrodes (electrical field), d is the distance between electrodes, E/d is the field strength, and q is the net charge of molecule.

If the molecule or particle with the charge q is placed in the electrical field E, it will move at a constant velocity v, which is determined by the balance between the electric force Eq and the viscous drag fv. The equation that represents this phenomenon is

$$Eq = fv$$

where Eq is the electric force, f is the frictional coefficient of the molecule (a function of the physical parameters of that molecule), v is the charge velocity, and fv is viscous drag.

Using this equation, it is possible to determine the characteristic *mobility* u of a defined particle, which is its velocity for a given external electrical field:

$$\text{Mobility} = u = v/E = q/f$$

The term *electrophoresis* refers to the transport of particles through a solvent by an electric field. If a charged molecule is placed in a vacuum, it will accelerate until it finally collides with the electrode. This does not occur, however, if the molecule is in solution or in a matrix. In these instances, the force of the electrical field is opposed by the friction that occurs between the accelerating molecule and the solution. The degree of the drag on this molecule is therefore dependent on the *size* and *shape* of the molecule as well as the *viscosity* of the medium through which it moves. If electrophoresis is carried out in a solution such as sucrose, the extent of this drag can be calculated using the Stokes equation:

$$F = 6\pi r \eta v$$

where F is the drag force exerted on spherical molecule, r is the radius of a spherical molecule, η is the solution viscosity, and v is the velocity at which the molecule is moving.

It is thus possible to estimate the velocity at which a molecule moves in an electric field in a solution environment:

$$v = Eq/d6\pi r\eta$$

where v is velocity.

This equation demonstrates that the movement of the molecule is proportional to the field strength and charge on the molecule but is inversely proportional to its size and solution viscosity.

A major complication in these theories is that a charged particle suspended in an electrolyte is surrounded by an ion atmosphere that shields it from the applied field. This is because the charged molecule attracts ions. Since the ionic atmosphere is disrupted by the applied field as well as the motion of the particle through the medium, many of the theories of electrophoresis have failed to provide detailed information about macromolecular structures (Freifelder, 1976). Furthermore, while these simplified equations make it possible to appreciate the movement of small defined molecules or particles in an electrophoretic field, the movement of complex macromolecules in electrophoretic fields and in the presence of rigid matrices such as polyacrylamide becomes much more complicated. The complexities of protein composition will also make precise predictions of electrophoretic mobility difficult.

Proteins possess charges as a result of the basic (arginine and lysine) and acidic (glutamic and aspartic acids) amino acids. Complex proteins that have been modified due to posttranslational modification such as glycosylation, sulfation, and phosphorylation have additional charges. Since these proteins contain a net charge, they may migrate in an electrical field at any pH other than the isoelectric point of the protein itself. Their rate of migration in this electrophoretic field will also depend on the charge density, that is, the ratio of charge to mass. The greater the ratio of charge to mass, the faster the molecule will migrate.

IV. CHOICE OF BUFFERS FOR ELECTROPHORESIS

To date, a great deal of time and effort has been spent on determining which buffers are optimal for carrying out electrophoretic analyses (see reviews by Chrambach and Rodbard, 1981; Hames and Rickwood, 1981).

A buffer system should be chosen because it (1) maintains a constant pH within the reservoirs as well as within matrices such as poly-acrylamide, and (2) acts as an acceptable electrolyte conducting current across the electric field. A buffer should further be chosen that does not interact with molecules that are being separated because this might alter the rate at which a molecule migrates during electrophoresis and give artifactual results (Cooper, 1977). Not only is the composition of the buffer important in electrophoresis, but the ionic strength and concentration of the buffer is critical. If the electrolyte concentration is too low, the migrating macromolecules that are being separated will conduct a large portion of the current and will therefore not be resolved adequately. If the electrolyte is too high in concentration, the amount of current conducted will increase while the voltage decreases. This will lead to the generation of heat as well as a decrease in macromolecular migration.

In theory, electrophoresis can be carried out between pH 2.5 and 11, but in practice the limits are pH 3–10, since reactions such as deamidation can occur at extremes of pH. If nondissociating electrophoresis is to be carried out in order to separate molecules such as proteins according to their size and charge density, the choice of a buffering system is critical.

The further the pH of the buffer is from the isoelectric points of the proteins to be separated, the higher the charge on the proteins. The relative difference will alter the time required for electrophoretic separation of proteins. If electrophoresis is carried out in the presence of such reagents as the anionic detergent sodium dodecyl sulfate (SDS), the exact range of pH during electrophoresis is not as critical because the detergent–polypeptide complexes will all be negatively charged. The pH ranges of this system are only important if a "stacking" or protein concentration–discontinuous buffer system is used (Hames, 1981).

V. BASIC TYPES OF ELECTROPHORESIS

The three major types of electrophoresis include moving boundary, zone, and continuous. The types of electrophoresis commonly used involve variations of these types of electrophoresis.

A. Moving Boundary Electrophoresis

This type of electrophoresis has properties similar to boundary sedimentation in that it is an analytical method that has been used primarily for the determination of mobilities and isoelectric points of proteins. This method was developed by the early studies of Tiselius and associates

(Tiselius, 1937; Tiselius and Kabat, 1939), in which serum was separated into α-, β-, and γ-globulins and albumin. This method was summarized in detail by Garvey *et al.* (1977) and Freifelder (1976). The apparatus used is a compartmented glass cell in which one section is filled with protein sample (dialyzed against the same buffer as the other section). The cell is connected through a closed system containing the same buffer to reversible silver–silver chloride (Ag–AgCl) electrodes. (When an electric potential is applied to the system, the protein molecules will migrate at a rate that is dependent on their charge, size, and field strength.) The position of the molecules suspended in the solution as a function of time is determined by schlieren optics. (The schlieren pattern is obtained when light passing through regions of varying concentration is deviated because of the change in the index of refraction, which varies with concentration.)

The direct relationship between the refractive index and the protein concentration permits quantitative evaluation of the relative percentage of each protein component. Because alternative methods are now available that are more convenient and efficient, this method is seldom used as an analytical tool for studying proteins.

B. Zone Electrophoresis

This type of electrophoresis involves the application of macromolecules in a narrow zone or band at a suitable distance from electrodes such that during electrophoresis, proteins of different mobilities travel as discrete zones that gradually separate from each other as electrophoresis proceeds (Hames, 1981). This method differs from moving boundary electrophoresis in several ways (Garvey *et al.*, 1977): (1) the macromolecules migrate as separate zones rather than as advancing boundaries of overlapping zones; (2) the macromolecules are stabilized against convection because their movement through water is limited by pores of support matrices; and (3) components are not detectable until the end of the experiment, at which time the appropriate detection method is used.

Zone electrophoresis can be carried out in free solution but is more practically carried out in a support medium such as starch or polyacrylamide. This method is commonly used to analyze complex protein mixtures.

1. Zone Electrophoresis with Continuous Buffer Systems

When electrophoresis is carried out using the same buffer ions at constant pH throughout the sample, gel, and electrode reservoirs (even if differing concentrations are used), the system is termed continuous. Because resolving gels generally have small pore sizes, "molecular sieving" or separation of proteins according to size will occur during electrophoresis (Hames, 1981).

2. Zone Electrophoresis with Discontinuous Buffer Systems

Discontinuous buffer systems are also known as multiphasic buffer systems; they use different buffer ions in the gel as compared with those in the electrode reservoirs. These may differ in both pH and buffer composition as well. One popular system that is used includes the use of a "stacking" gel, which offers the advantage that relatively large volumes of dilute protein samples can be applied to gels with good resolution of sample components (Ornstein, 1964; Davis, 1964; Hames, 1981).

3. Zone Electrophoresis with Dissociating Buffer Systems

The majority of electrophoretic studies now employ zone electrophoresis in the presence of detergents such as SDS or urea in the buffer system designed to dissociate all proteins into their individual polypeptide subunits (see Section VIII and Chapter 3). While sodium dodecyl sulfate–polyacrylamide gel electrophoresis (SDS–PAGE) has become the most widely used method for separation of proteins by electrophoresis, this method is limited in that the biological activities of proteins are generally destroyed by denaturing reagents, unlike buffer systems in which non-dissociating conditions are used.

C. Continuous Electrophoresis

This type of electrophoresis is similar to zone electrophoresis except that it is primarily a preparative scale procedure that is achieved by the continuous addition of sample to a zone (Freifelder, 1976). While this methodology is not commonly used, the continuing advances in equipment in this area should make these methodologies more popular in the future.

VI. MOVEMENT OF MOLECULES IN SUPPORT MEDIA DURING ELECTROPHORESIS

Over the years, a variety of matrices have been used as supporting material for separation of molecules during electrophoresis. Among the most common supports are paper, cellulose acetate, starch, agarose, and polyacrylamide. These supports will affect the mobility of charged molecules in many ways including adsorption to the medium (such as to the cellulose hydroxyl groups of cellulose paper) or reduced movement due to "molecular sieving," as is the case with PAGE. Since the interactions of complex molecules with such matrices may alter their mobility dramatically during electrophoresis, it becomes extremely difficult to predict mathematically the theoretical mobility of a given macromolecule such as glycoprotein.

VII. BASIC PRINCIPLES OF ISOELECTRIC FOCUSING

A. Definition of Isoelectric Point

Proteins are composed of different amounts of amino acids R, many of which have side groups that become charged depending on the pH of the surrounding media. For example, carboxyl groups (COOH) have a zero net charge at low pH and a negative net charge at high pH:

$$R\text{-}COO^- + H^+ \rightarrow R\text{-}COOH \qquad \text{(at low pH)}$$

$$R\text{-}COOH + HO^- \rightarrow R\text{-}COO^- + H_2O \qquad \text{(at high pH)}$$

Amino groups (lysine residues), imidazolium groups (histidine residues), and guanidinium groups (arginine residues) bind a proton at low pH, causing them to become positively charged; but at high pH, these groups will dissociate to leave a group with a zero net charge:

$$R\text{-}NH_2 + H^+ \rightarrow R\text{-}NH_3^+ \qquad \text{(at low pH)}$$

$$R\text{-}NH_3^+ + OH^- \rightarrow R\text{-}NH_2 + H_2O \qquad \text{(at high pH)}$$

The net charge on any protein will therefore be due to the sum of the positive and negative charges on all amino acid residues as well as other side groups such as carbohydrates, phosphates, and sulfates. At one specific pH, a protein will have no net charge (i.e., the molecule will have an equal number of positive and negative charges). This pH level is referred to as the isoelectric point (pI) for that protein (see discussion by Vesterberg, 1976). The term pI in isoelectric focusing (IEF) refers to the pH region reached (or "focused") by the protein or ampholyte component after an arbitrary period of focusing in a gel or other anticonvection medium of arbitrary composition and at an arbitrary temperature, making comparison of results among laboratories difficult (Chrambach and Baumann, 1976).

Initial studies in IEF were based on "focusing" molecules to an "equilibrium" state in sucrose density-gradient experiments in which (at relatively low potential gradients), a linear pH gradient is developed. Proteins then focused into bands of constant pI after 1 day of operation (Svensson, 1961). It has been pointed out, however, that the term "equilibrium," which denotes the final state of IEF, is ill chosen (Chrambach and Baumann, 1976). As these investigators have pointed out, the term "equilibrium" belongs to the field of thermodynamics and has a specific meaning that describes the perfect balance between forward and backward reactions. Isoelectric focusing does not fit this criterion, however, because it involves at least the continuous transport of proteins and hydroxyl ions at a rate sufficient to produce a small but finite current flow. Therefore, since there is a net transport, there cannot be equilibrium. The proper term for the "final" state in IEF is therefore *steady state* (Catsimpoolas, 1973; Chrambach and Baumann, 1976). Furthermore, it has been observed that all amphoteric electrolytes, carrier ampholytes, as well as proteins approach steady state at characteristic individual rates and that the steady state for each species is both limited and varied. Because of these variables, it is becoming increasingly critical that laboratories using these methods standardize their IEF systems (see discussion in Chapter 11).

B. Definition of Ampholytes

If proteins are to be separated by electrofocusing, it is necessary to establish a stable linear gradient, where possible. *Carrier* ampholytes are molecules used to generate a pH gradient during electrofocusing (Vesterberg, 1976; Rilbe, 1976). (Ampholine is the trademark used by LKB for their carrier ampholytes.) Ampholytes commonly used are those described by Vesterberg (1976), which are composed of isomers and homologues of aliphatic polyaminopolycarboxylic acids with pH values ranging from 2.5 to 11. Some commercially available ampholyte mixtures may contain hundreds of such molecular species. The most desirable properties for carrier ampholytes include (1) even conductivity at the isoelectric point, (2) high buffering capacity, (3) solubility at the isoelectric point, and (4) minimum interaction with focused proteins (LKB Product Manual, 1979).

1. Even Conductivity

In order to obtain a flow of current between electrodes, adequate conductivity is required. Furthermore, the production of even field strength and the prevention of local high resistances where overheating may occur is essential (see discussion on polyacrylamide gel breakage in Chapter 14).

2. High Buffering Capacity

Ampholytes should have adequate buffering capacity to limit the extent of local alterations of pH in any part of the pH gradient by the proteins that will be focused there.

3. Solubility at Isoelectric Point

Many proteins will precipitate when they reach their isoelectric point. This is an undesirable property of ampholytes because precipitation reactions may interfere with migration of other molecules in a poly-acrylamide matrix.

4. Minimum Interaction with Focused Proteins

Staining reactions as well as electroelution and antibody binding reactions depend on the removal of molecules that might inhibit binding of dyes, antibodies, and so forth. It is therefore critical to use ampholytes that do not bind to proteins. While carrier ampholytes (such as LKB Ampholine) have only amino and carboxyl functional groups that interact minimally with proteins, there have been some reports of ampholyte binding to such charged molecules as heparin (Righetti and Gianazza, 1978) and an acidic phosphoprotein (Jonsson et al., 1978).

C. Experimental Parameters Affecting Measurement to Apparent Isoelectric Points of Proteins

Experimentally, many factors can affect measurement of the apparent pI of the proteins during isoelectrophoresis. These must be taken into account if one is to interpret protein patterns adequately (see reviews by Gelsema and DeLigny, 1977; LKB Product Manual, 1979; An der Lan and Chrambach, 1981).

1. Protein Effects

Since proteins have their own inherent charge, they can act as ampholytes themselves and affect the pH during electrofocusing. The amount of protein applied to gels in proportion to the amount of ampholytes is therefore important. In some instances, a purified protein will not migrate to the same region of the IEF gel as it does when it is focused in the presence of a complex protein mixture.

2. Ampholyte Gradient Formation

Generally, carrier ampholytes have a much higher mobility than do proteins. The gradient formation of ampholytes therefore occurs at a faster rate than the rate at which proteins reach their steady state. Some proteins will reach their steady point in a short time, while other proteins (generally ones of greater molecular weights) may never reach steady state due to matrix restrictions. Because different sources of ampholytes can vary as well as different lots of ampholytes, the pH gradient will frequently vary from experiment to experiment. The use of internal charge standards therefore becomes critical for comparing pH gradients.

3. Effect of Time and Voltage

The number of volt-hours (V-hr) used during any experimental IEF run will dramatically affect the results of an experiment. Because some proteins may reach their steady-state pH position more rapidly than others, variation of times of electrofocusing as well as voltages used for focusing may be necessary. If complex protein samples are being analyzed, the patterns of protein distributions vary. Furthermore, one may obtain varied results even if the same number of total volt-hours is used, but different combinations of these two parameters are used (e.g., 700 V $\times$ 15 hr = 10,500 V-hr versus 500 V for 21 hr = 10,500 V-hr). Again, the use of internal standards for the determination of optimal resolution in focusing systems is important.

4. Molecular Interactions

Ampholyte–protein interactions may alter pI values and further interfere with protein staining.

5. Temperature Effects

Temperature is another factor that may influence the apparent pI of a protein. This effect is particularly pronounced if the temperature varies in different parts of the gel. Many investigators will measure the pH of their polyacrylamide gel slices after focusing. If there is a temperature difference in the gels at the time the pH is measured, this method can give inaccurate information concerning the actual pH of the gel. In general, as the temperature rises, the apparent pI will fall.

6. Alkaline Proteins

It is generally more difficult to carry out focusing of alkaline proteins. The alkaline carrier ampholytes of many commercially available ampholyte mixtures have uneven conductivity profiles. (Methods to analyze basic proteins have been developed and are discussed in Section E.)

7. Solvent Effects

The addition of a variety of solvents or solubilization reagents to gels or samples can further alter the relative pI of a protein. For example, high concentrations of urea can change the pI by several tenths of a pH unit.

8. Ionic Effect

Many nonamphoteric ions may also have an effect on the apparent pI of proteins. For example, atmospheric CO_2 will dissolve rapidly in alkaline solutions, resulting in the reduction of the pH of the solution. This can most easily be controlled by thoroughly degassing solutions prior to polymerization of gels and of electrode buffers just prior to electrofocusing.

9. Salt Effect

Excess salt in either the polymerization mixture or the sample can also affect the pH gradient. Samples prepared in salt solutions or samples concentrated by salt precipitation (e.g., ammonium sulfate) should therefore be desalted by dialysis or chromatography prior to electrofocusing, if this is likely to be a problem.

10. Sample Preparation

Sample preparation is discussed in more detail in Chapter 3, but it is important to be aware of conditions that can dramatically alter the apparent pI of a protein. Many labeling methods such as iodination with iodoacetic acid may alter the charge of a protein and should be avoided if one wishes to determine the relative pI of a protein. Heating in urea may cause carbamylation of proteins, which can dramatically alter the pI of a protein and should therefore be avoided. Other treatments such as precipitation with acids may alter protein structure and should be avoided if possible.

D. Determination of pH in Isoelectric Focusing Gels

To date, a variety of methods have been used to measure the pH gradients in IEF gels. One of the most popular methods has been to slice a gel into segments, followed by diffusion of ampholytes and buffers into water. The pH of that solution is then measured. A second method has been the use of electrodes designed to measure pH directly in gel matrices. Both measurements can give erroneous estimations of pH values for a variety of reasons. Temperature during focusing may be different from the temperature at which the pH is determined, and these methods generally do not account for protein effects or molecular interactions in gels. For these reasons, it is critical that the investigator not overemphasize pH measurements until all these factors are taken into consideration. The use of internal IEF standards has proved the most reliable method for comparison of protein patterns among different gels as well as among different laboratories (N. L. Anderson and Hickman, 1979; see Chapter 3, Fig. 3.3).

E. Equilibrium versus Nonequilibrium Isoelectric Focusing

In reality, it is highly unlikely that all proteins in a complex protein mixture will focus to steady state during electrofocusing. It is common for the pH gradient at the cathodic end to break down, giving rise to cathodic drift. Because of this problem, most basic proteins will not focus in discrete spots at the cathodic end of the gels. A gel system referred to as nonequilibrium gel electrophoresis (NEPHGE) has been developed to resolve basic as well as acidic proteins (P. H. O'Farrell et al., 1977). (Again, it is important to note that the term "equilibrium" is not an accurate term for IEF as described above.)

VIII. BASIC PRINCIPLES OF POLYACRYLAMIDE GEL ELECTROPHORESIS

A. Acrylamide Gel Matrix

The optimal matrix for carrying out protein separation by electrophoresis is one that is stable, that decreases or eliminates convection, and that does not react with the sample or retard its movement. One of the most widely used matrices is polyacrylamide. This matrix is composed of acrylamide copolymerized with N,N'-methylene-bis(acrylamide) in the presence of free radicals (Richards and Lecanidou, 1974) (Fig. 1.2).

The polymerization of acrylamide occurs when ammonium persulfate is dissolved in H_2O, so that free sulfate radicals are formed. When these free radicals come into contact with acrylamide, long polymer chains of acrylamide form. Tetramethylenediamine (TEMED) is used in the polymerization procedure to serve as a catalyst of gel formation because of its ability to exist in a free radical form. The acrylamide polymer chains will not gel except in the presence of N,N'-methylene-bis(acrylamide), which will form the crosslinks between the long acrylamide polymer chains (see discussion by Cooper, 1977; Chrambach and Rodbard, 1981).

B. Role of Catalysts (Initiators) in Acrylamide Polymerization

Catalysts for PAGE should be selected for their reproducibility and efficiency in gel polymerization. Gel polymerization is commonly initiated chemically by ammonium persulfate and TEMED or by the photochemical method using riboflavin. The riboflavin method is not commonly used because it requires special equipment for photopolymerization as well as several hours for complete polymerization to occur (Righetti *et al.*, 1981). When polymerization is initiated, the ammonium persulfate yields a per-

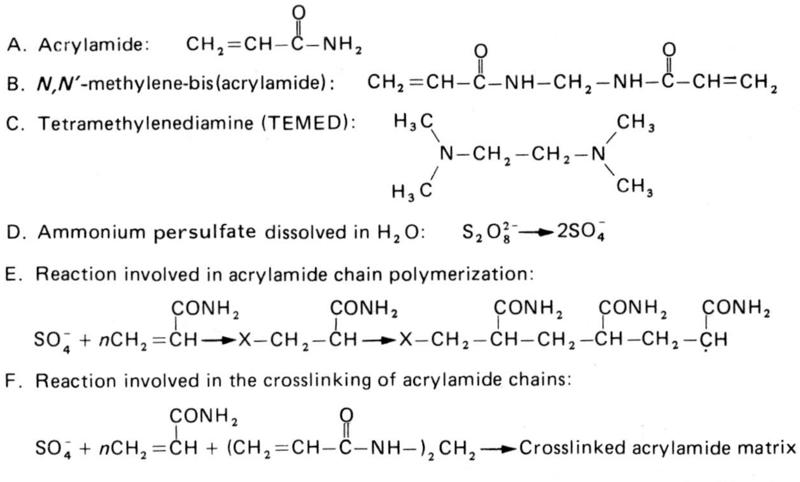

A. Acrylamide: $CH_2{=}CH{-}\overset{\overset{\displaystyle O}{\|}}{C}{-}NH_2$

B. N,N'-methylene-bis(acrylamide): $CH_2{=}CH{-}\overset{\overset{\displaystyle O}{\|}}{C}{-}NH{-}CH_2{-}NH{-}\overset{\overset{\displaystyle O}{\|}}{C}{-}CH{=}CH_2$

C. Tetramethylenediamine (TEMED): $\begin{array}{c}H_3C\\ \diagdown \\ \quad N{-}CH_2{-}CH_2{-}N \\ \diagup \\ H_3C\end{array}\begin{array}{c}CH_3 \\ \diagup \\ \\ \diagdown \\ CH_3\end{array}$

D. Ammonium persulfate dissolved in H_2O: $S_2O_8^{2-} \longrightarrow 2SO_4^-$

E. Reaction involved in acrylamide chain polymerization:

$SO_4^- + nCH_2{=}\overset{\overset{\displaystyle CONH_2}{|}}{CH} \longrightarrow X{-}CH_2{-}\overset{\overset{\displaystyle CONH_2}{|}}{CH} \longrightarrow X{-}CH_2{-}\overset{\overset{\displaystyle CONH_2}{|}}{CH}{-}CH_2{-}\overset{\overset{\displaystyle CONH_2}{|}}{CH}{-}CH_2{-}\overset{\overset{\displaystyle CONH_2}{|}}{CH}$

F. Reaction involved in the crosslinking of acrylamide chains:

$SO_4^- + nCH_2{=}\overset{\overset{\displaystyle CONH_2}{|}}{CH} + (CH_2{=}CH{-}\overset{\overset{\displaystyle O}{\|}}{C}{-}NH{-})_2CH_2 \longrightarrow$ Crosslinked acrylamide matrix

Figure 1.2. Chemical structures of acrylamide and reactions involved in acrylamide gel polymerization. (Modified from Cooper, 1977.)

sulfate free radical that activates the TEMED. The TEMED therefore acts as an electron carrier that provides an unpaired electron and converts the acrylamide monomer to a free radical state. This activated monomer in turn reacts with unactivated monomer to begin the chain elongation. In the presence of bis(acrylamide), the elongating polymer chains are crosslinked, resulting in matrix formation with random pore sizes (see discussions by Cooper, 1977; BioRad Laboratories, 1984). The average pore size will therefore depend on the concentrations of acrylamide, bis(acrylamide), and catalysts, as well as the conditions used for polymerization.

The rate of the polymerization will depend on the type and concentration of the initiators. In general, visible gelation of polyacrylamide should occur within 10–20 min (Chrambach and Rodbard, 1981; BioRad Laboratories, 1984). The concentrations of initiators will affect the average polymer chain length and therefore may alter the mechanical stability of the gel (Chrambach and Rodbard, 1981). Excess amounts of ammonium persulfate and TEMED may also interact with proteins, increase the buffer pH, and therefore alter the banding patterns of proteins in gels (Dirksen and Chrambach, 1972; Chrambach *et al.*, 1976; Gelfi and Righetti, 1981).

In general, catalysts concentrations may have to be altered if the polymerization rate is too fast or too slow. [*Note:* the catalyst concentrations needed for optimal polymerization usually decrease with increasing gel concentrations (Chrambach and Rodbard, 1981).] Because TEMED is subject to oxidation and both TEMED and ammonium persulfate are hygroscopic, the accumulation of water in these reagents may cause inactivity. It is therefore critical that high-quality reagents be used and that the reagents be made fresh at every use. (If ammonium persulfate is made fresh and frozen immediately in small aliquots, it should remain active for several months.)

C. Acrylamide Pore Size

The pore size of the polymerized acrylamide will depend on (1) the amount of acrylamide used per unit volume of reaction medium; and (2) the degree of crosslinking. The ratio of acrylamide to bis(acrylamide) is therefore critical in the formation of pore size. (*Note:* Pore formation is random, therefore, pore sizes will never be totally uniform.) The average pore size will depend on the ratio of the percentage weight of acrylamide to the total gel volume. For more details, refer to Cooper (1977), and Chrambach and Rodbard (1981).

If PAGE is being carried out to obtain maximal resolution of protein (especially if nondissociating gel systems are being used), the optimal concentrations of acrylamide will have to be determined. This can be carried out by measuring the mobility of each protein in a series of gels of different acrylamide concentrations and then by constructing a Ferguson plot for each protein of interest (Rodbard and Chrambach, 1974; Chrambach and Rodbard, 1981). This is a plot of the $\log_{10}$ relative mobility (Rf) versus the gel concentration, %T (percentage of total monomer including grams of acrylamide plus bisacrylamide in 100 ml). The relative mobility of a protein will therefore depend on the mobility of the protein of interest relative to the mobility of a tracking dye marker. Although this system has been used to estimate the molecular size of native proteins (Hedrick and Smith, 1968; Chrambach and Rodbard, 1981), this measurement is only valid if standard proteins used to generate the calibration curves have the same shape with the same degree of hydration and partial specific volume as the proteins of interest (Hames, 1981). It has therefore become popular to estimate molecular weights of polypeptides using PAGE in the presence of SDS after reduction of protein disulfide bonds with a thiol reagent because the differences in charge density and conformation among proteins can reduce some experimental variability (Hames, 1981).

D. Effect of Acrylamide Pore Size on Separation of Proteins

Acrylamide gels used in IEF generally have a greater ratio of bis-(acrylamide) to acrylamide (1.8 : 30) than do SDS–PAGE gels (0.8 : 30); therefore, the pores will be larger. Since proteins are separated according to the net charge during IEF, it is important that there be a minimum retention in the acrylamide pores. The second-dimension separation, however, is based on the relative molecular weight of the proteins. A gradient of acrylamide pore sizes is therefore prepared using different percentages of total acrylamide to maximize the resolution of separation of proteins. The use of gradient gels will give a separation of a wider range of molecular weights than will nongradient gels as well as improve the resolution of protein separation.

IX. PRINCIPLES OF A STACKING GEL

The use of stacking gels in electrophoresis in polyacrylamide gels has become popular because these methods have been shown to increase the resolution in concentrating proteins in gels in order to increase resolution of protein separation. While this is a useful technique, it is important to note that the use of high-resolution gradient gels will frequently give optimal resolution due to molecular sieving effects of varying pore sizes without the use of a stacking gel. For those investigators who use this method, it is, however, important to understand the principles of protein separation when using a stacking gel.

The principal electrophoresis method that employs a stacking gel is that of discontinuous pH, or disc, electrophoresis, which is a modification of zone electrophoresis systems. In this gel system, the lower running gel is the gel in which the proteins separate (usually 5–20% acrylamide or acrylamide gradient). The stacking gel is a smaller layer of acrylamide (2–4%), which gives larger pore size. The buffer system of this gel will cause proteins to migrate rapidly and pile up in the stacking layer before entering the running layer. The principle of the stacking gel as described by Cooper (1977) is as follows:

1. The buffer used in the stacking gel is an amino acid such as Tris, which has its pH adjusted with HCl to 6.5, which is 2 pH units below the Tris buffer in the running gel, which is 8.7.
2. The buffer used in the protein sample is the same as that used for the stacking gel.
3. The buffer for the upper reservoir is also an amine (Tris) that is adjusted to the same or slightly higher pH as the running gel with a weak acid such as glycine.
4. Glycine will exist in the upper buffer as both a zwitterion with a net charge of 0 and a glycinate ion with a charge of -1 (a *zwitterion* is a dipolar ion that contains both a positive and a negative charge but is neutral as a whole):

$$\text{pH 8.7:} \quad {}^+NH_3CH_2COO^- \rightarrow NH_2CH_2COO^- + H^+$$

5. When glycinate anions enter a low pH in the sample buffer along with the protein, chloride, and bromophenol blue, they shift their equilibrium to the immobile zwitterion form:

$$\text{pH 6.5:} \quad NH_3CH_2COO^- + H^+ \rightarrow {}^+NH_2CH_2COO^-$$

6. Failure of glycine zwitterions to move into the sample and stacking gel creates a deficiency of mobile ions and therefore decreases the current flow.
7. Because a constant current must be kept throughout the entire

electric system, the area between the leading chlorine ions and the trailing glycinate ions forms a localized high-voltage gradient.

8. In this strong electric field, the anionic proteins migrate rapidly and stack in the gel containing large pores that minimally impede protein migration.

9. When proteins overtake leading chlorine ions, they slow down because there is no ion deficiency where these are present, hence no large field strength.

10. All proteins therefore pile up to form a tight disc between the glycinate and chloride ions.

11. As the disc of proteins moves into the running gel, the migration is slowed down by the smaller pores of the acrylamide, permitting the glycinate ions to catch up with the proteins.

12. When glycinate ions move into running gel buffer (pH 8.7), they become fully charged; from this point, there is no more ion deficiency, but a constant field strength throughout the gel.

X. SODIUM DODECYL SULFATE–POLYACRYLAMIDE GEL ELECTROPHORESIS

The anionic detergent, sodium dodecyl (laurel) sulfate, $NaDodSO_4$, $(H_3(CH_2)_{10}CH_2OSO_3Na)$ is now used routinely in PAGE. SDS will bind to the hydrophobic regions of proteins and separate many of them into component subunits. It will further give a large negative charge to the denatured, randomly coiled polypeptides so that they can easily be separated using electrophoresis. Weber and Osborn (1969) and Laemmli (1970) used this method to determine the mobilities of a variety of proteins in polyacrylamide gels in the presence of SDS and found that the mobilities of these proteins were a linear function of the logarithms of their molecular weight. Unfortunately, this method cannot be used to determine the molecular weights of all proteins. For example, glycoproteins with high carbohydrate content have been found to migrate in this type of electrophoresis at slower rates than would be expected from their molecular mass. The low mobility of these carbohydrate-containing proteins has been shown to be the result of their binding a smaller amount of SDS on a weight basis than is true of standard proteins of the same mass. The applications of this method are covered in detail throughout this text. The combination of protein separation according to charge by IEF in conjunction with SDS–PAGE has provided the most effective procedure for analyzing complex protein samples.

XI. HIGH-RESOLUTION TWO-DIMENSIONAL POLYACRYLAMIDE GEL ELECTROPHORESIS

Historically, many different methods have been developed to analyze and purify proteins. Combinations of many different methods have also been used to improve the resolution of protein separations. In fact, several early two-dimensional electrophoretic separation methods have been described (Smithies and Poulik, 1956; Dale and Latner, 1969; Stegemann, 1970; Martini and Gould, 1971; see also review by N. G. Anderson and Anderson, 1982).

The methodology that has recently become popular for use in protein separation by O'Farrell (1975) has been termed high-resolution two-dimensional polyacrylamide gel electrophoresis. This method involves the separation of proteins in the first dimension according to their charge

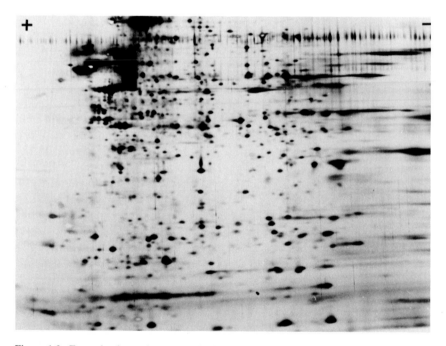

Figure 1.3. Example of protein pattern obtained by separation of proteins by high-resolution two-dimensional PAGE and detected with the color-based silver stain (Gelcode). (1) Proteins separated in the first dimension by IEF. (2) Proteins separated in the second dimension by SDS–PAGE (10–20% acrylamide gradient). (*Note:* This photograph represents only a portion of a two-dimensional gel.)

and in the second dimension according to their relative mobility in SDS–PAGE (Fig. 1.3). This method was made practical by the work of O'Farrell (1975); subsequently, equipment and methods have been improved such that large numbers of analyses can be easily and reproducibly performed (N. G. Anderson and Anderson, 1978; N. L. Anderson and Anderson, 1978).

REFERENCES

An der Lan, B., and Chrambach, A., 1981, Analytical and preparative electrofocusing, in: *Gel Electrophoresis of Proteins: A Practical Approach* (B. D. Hames and D. Rickwood, eds.), pp. 157–185, IRL Press, Oxford, England.

Anderson, N. G., and Anderson, N. L., 1978, Analytical techniques for cell fractions. XXI. Two-dimensional analysis of serum and tissue proteins: Multiple isoelectric focusing, *Anal. Biochem.* **85**:331–340.

Anderson, N. G., and Anderson, N. L., 1982, The human protein index, *Clin. Chem.* **28**:739–748.

Anderson, N. L., and Anderson, N. G., 1978, Analytical techniques for cell fractions. XXII. Two-dimensional analysis of serum and tissue protein: Multiple gradient-slab electrophoresis, *Anal. Biochem.* **85**:341–354.

Anderson, N. L., and Hickman, B. J., 1979, Analytical techniques for cell fractions. XXIV. Isoelectric point standards for two-dimensional electrophoresis, *Anal. Biochem.* **93**:312–320.

BioRad Laboratories, 1984, Acrylamide Polymerization—A Practical Approach, Bulletin #1156.

Catsimpoolas, N., 1973, Immuno-isoelectrofocusing, *Ann. N.Y. Acad. Sci.* **209**: 144–146.

Chrambach, A., and Baumann, G., 1976, Isoelectric focusing in polyacrylamide gel, in *Isoelectric Focusing* (N. Catsimpoolas, ed.), pp. 77–91, Academic Press, New York.

Chrambach, A., and Rodbard, D., 1981, "Quantitative" and preparative polyacrylamide gel electrophoresis, in: *Gel Electrophoresis of Proteins: A Practical Approach* (B. D. Hames and D. Rickwood, eds.), pp. 93–141, IRL Press, Oxford, England.

Chrambach, A., Jovin, T. M., Svendsen, P. J., and Rodbard, D., 1976, in: *Methods of Protein Separations,* Vol. 2 (N. Catsimpoolas, ed.), pp. 27–144, Plenum Press, New York.

Cooper, T. G., 1977, Electrophoresis, in: *The Tools of Biochemistry* (T. G. Cooper, ed.), pp. 194–211, John Wiley & Sons, New York.

Dale, G., and Latner, A. L., 1969, Isoelectric focusing of serum proteins in acrylamide gels followed by electrophoresis, *Clin. Chim. Acta* **24**:61–68.

Davis, B. J., 1964, Disc electrophoresis. II. Method and application to human serum proteins, *Ann. NY Acad. Sci.* **121**:404–427.

Dirksen, M. L., and Crambach, A., 1972, Studies on the Redox state in polyacrylamide gels, *Separation Sci.* **7**:747–772.

Faraday, M., 1834, Experimental researches in electricity—seventh series, *Philos. Trans. Roy. Soc.* **124**:77–122.

Freifelder, D., 1976, Electrophoresis, in: *Physical Biochemistry: Applications to Biochem-*

istry and Molecular Biology (C. I. Davern, ed.), pp. 211–232, W. H. Freeman, San Francisco.

Garvey, J. S., Cremer, N. E., and Sussdorf, D. H., 1977, Electrophoresis, in: *Methods in Immunology,* 3rd ed. (J. S. Garvey, N. E. Cremer, and D. H. Sussdorf, eds.), pp. 93–97, W. A. Benjamin, Inc., Reading, Massachusetts.

Gelfi, C., and Righetti, P. G., 1981, Polymerization kinetics of polyacrylamide gels. I. Effect of different crosslinkers, *Electrophoresis* 2:213–219.

Gelsema, W. J., and DeLigny, C. L., 1977, Isoelectric focusing as a method for the characterization of ampholytes, *J. Chromatogr.* 130:41–50.

Hames, B. D., and Rickwood, D., 1981, *Gel Electrophoresis of Proteins: A Practical Approach,* IRL Press, Oxford, England.

Hames, B. D., 1981, An introduction to polyacrylamide gel electrophoresis, in: *Gel Electrophoresis of Proteins: A Practical Approach* (B. D. Hames and D. Rickwood, eds.), pp. 1–86, IRL Press, Oxford, England.

Hedrick, J. L., and Smith, A. J., 1968, Size and charge isomer separation and estimation of molecular weights of proteins by disc gel electrophoresis, *Arch. Biochem. Biophys.* 129:155–164.

Jonsson, M., Fredriksson, S., Jontell, M., and Linde, A., 1978, Isoelectric focusing of the phosphorylation of rat-incisor dentin in ampholine and acid pH gradients, *J. Chrom.* 157:235–242.

LKB Product Manual, 1979, *Electrofocusing Seminar Notes,* Part 1: *Basic Principles,* LKB Produkter, AB, Bromma, Sweden.

Laemmli, U. K., 1970, Cleavage of structural proteins during the assembly of the head of bacteriophage T4, *Nature (Lond.)* 277:680–688.

Martini, O. H. W., and Gould, H. J., 1971, Enumeration of rabbit reticulocyte ribosomal proteins, *J. Mol. Biol.* 62:403–405.

Murphy, D. B., and Rousseau, V., 1969, *Foundations of College Chemistry,* Ronald Press, New York.

O'Farrell, P. H., 1975, High-resolution two-dimensional electrophoresis of proteins, *J. Biol. Chem.* 250:4007–4021.

O'Farrell, P. Z., Goodman, H. M., and O'Farrell, P. H., 1977, High-resolution two-dimensional electrophoresis of basic as well as acidic proteins, *Cell* 12:1133–1142.

Ornstein, L., 1964, Disc electrophoresis. I. Background and theory, *Ann. NY Acad. Sci.* 121:321–349.

Richards, E. G., and Lecanidou, R., 1974, Polymerization kinetics and properties of polyacrylamide gels, in: *Electrophoresis and Isoelectric Focusing in Polyacrylamide Gels* (R. C. Allen and H. R. Maurer, eds.), pp. 16–22, Walter deGruyter, Berlin.

Righetti, P. G., and Gianazza, E., 1978, Isoelectric focusing of heparin. Evidence for complexing with carrier ampholytes, *Biochim. Biophys. Acta* 532:137–146.

Righetti, P. G., Gelfi, C., and Bosisio, A. B., 1981, Polymerization kinetics of polyacrylamide gels. III. Effect of catalysts, *Electrophoresis* 2:291–295.

Rilbe, H., 1976, Theoretical aspects of steady-state isoelectric focusing, in: *Isoelectric Focusing* (N. Catsimpoolas, ed.), pp. 14–51, Academic Press, New York.

Rodbard, D., and Chrambach, A., 1974, Quantitative polyacrylamide gel electrophoresis: Mathematical and statistical analysis of data, in: *Electrophoresis and Isoelectric Focusing in Polyacrylamide Gels* (R. C. Allen and R. H. Maurer, eds.), pp. 28–62, Walter de Gruyter, Berlin.

Smithies, O., and Poulik, M. D., 1956, Two-dimensional electrophoresis of serum proteins, *Nature (Lond.)* 177:1033.

Stegemann, A., 1970, Protein mapping in polyacrylamide and its application to genetic analysis in plants, *Angew. Chem.* [in English] **9**:643.

Svensson, H., 1961, Isoelectric fractionation, analysis, and characterization of ampholytes in natural pH gradients. I. The differential equation of solute concentrations at a steady state and its solution for simple cases, *Acta Chem. Scand.* **15**:325–341.

Tiselius, A., 1937, A new apparatus for electrophoretic analysis of colloidal mixtures, *Trans. Faraday Soc.* **33**:524–531.

Tiselius, A., and Kabat, E. A., 1939, An electrophoretic study of immune sera and purified antibody preparations, *J. Exp. Med.* **69**:119–131.

Vesterberg, O., 1976, The carrier ampholytes, in: *Isoelectric Focusing* (N. Catsimpoolas, ed.), pp. 53–76, Academic Press, New York.

Weber, K., and Osborn, M., 1969, The reliability of molecular weight determinations by dodecyl sulfate polyacrylamide gel electrophoresis, *J. Biol. Chem.* **244**:4406–4412.

Applications and Strategies for Use of Polyacrylamide Electrophoresis

I. INTRODUCTION

The development of polyacrylamide gel matrices has resulted in the widespread use of electrophoresis for the characterization and purification of proteins. To date, most investigators employing these methods have used one-dimensional polyacrylamide gel electrophoresis (PAGE), either reducing or nonreducing, because these methods are reasonably simple and reproducible. Although these methods are adequate to answer some specific questions, the methods are limited and the results obtained are frequently overinterpreted. It is therefore critical to appreciate the strengths and limitations of each method.

II. NONREDUCING POLYACRYLAMIDE GEL ELECTROPHORESIS

The major reason for the use of gel systems that do not result in the fractionation of proteins into individual polypeptides or the reduction of disulfide bonds is that such treatments generally destroy biological activity. Such methods may also alter antigenic determinants; that is, they may disrupt a specific conformation necessary for antibody recognition (see discussion in Chapter 13). The obvious advantage of nonreducing PAGE is that some biochemical information can be obtained for a given protein without altering its biological activity. The main uses for this technology are therefore (1) to determine whether antibodies (primarily monoclonal) recognize conformational determinants, and (2) to detect and characterize proteins according to their biological activity (e.g., enzymatic activity) directly in polyacrylamide gels or from protein eluted from gels. Since a method for the estimation of molecular weights has been described that is based on the extent to which a protein migrates into acrylamide gels containing different total amounts of acrylamide (i.e., differing average pore sizes), this method can be used to determine the relative molecular weights of native proteins or to separate charge isomers of proteins (Hedrick and Smith, 1968). The main drawback to these nonreducing PAGE methods is that they are not high-resolution systems. Given the recent advantages in protein purification using high-pressure liquid chromatography (HPLC), affinity chromatography, and ion-exchange chromatography, it is now frequently easier to purify and chemically characterize native proteins by these higher-resolution methods than by nonreducing PAGE.

III. ONE-DIMENSIONAL SODIUM DODECYL SULFATE–POLYACRYLAMIDE GEL ELECTROPHORESIS

Since it was observed that mobilities of proteins in sodium dodecyl sulfate–polyacrylamide gel electrophoresis (SDS–PAGE) are a linear function of the logarithms of their molecular weights, this method and modifications of it (Laemmli, 1970) have been the most popular methods for protein analysis and characterization in all aspects of studies in biochemistry and cell biology as well as in clinical investigations.

This method has been widely applicable, since it is relatively straightforward, requires minimal equipment, and is generally reproducible from laboratory to laboratory. The major uses for this method are as follows:

1. To estimate the relative molecular weight of a protein (as long as the protein mixture applied to gel is not too complex and appropriate molecular-weight standards and acrylamide gradients are used)
2. To achieve a relative estimation of protein purity of a sample (*Note:* The sensitivity of this method is maximized if gels are overloaded and silver staining procedures are used.)
3. To screen antibodies using immunoblot methods (see Chapter 8 and Appendix 11)
4. To determine whether proteins are glycosylated by using lectins as probes in polyacrylamide gels or on proteins transferred to nitrocellulose (see Chapter 5 and Appendix 12).
5. To analyze proteins using peptide mapping (see Section V)

The major limitations of this method that have led to the overinterpretation of data generated from one-dimensional SDS–PAGE have been as follows:

1. The absolute assignment of molecular weights of proteins (see discussion on glycoproteins in Chapter 5)
2. The use of single-percentage nongradient polyacrylamide gels that do not resolve high- or low-molecular-weight proteins
3. The inability of this method to resolve proteins having the same molecular weight but different charge species

In general, high-resolution two-dimensional PAGE can be used to achieve more complete and accurate information than can one-dimensional PAGE, but the time, effort, and expertise required for this technology are greater.

IV. WHEN TO USE TWO-DIMENSIONAL POLYACRYLAMIDE GEL ELECTROPHORESIS

As a general rule, complex protein samples that can be analyzed by one-dimensional PAGE can be better resolved by high-resolution two-dimensional PAGE, which uses gradient acrylamide gels in the second dimension. If a laboratory is well equipped, these procedures can be carried out routinely and with reproducibility equivalent to that with one-dimensional SDS–PAGE methods. The advantages of using this methodology, in addition to or in place of one-dimensional PAGE methods, are considerable:

1. To achieve the best estimate of the degree of purity or complexity of a sample [When first analyzing a sample, it is advisable to run a wide-range ampholyte mix and to run the same sample in a one-dimensional lane next to the isoelectric focusing gel when running the second-dimension SDS gel (see Fig. 2.1). This will allow you to detect any proteins that may be extremely acidic or basic and that would therefore not be resolved by isoelectric focusing (IEF) unless special focusing conditions are used.]
2. To determine the degree of molecular weight and charge heterogeneity of proteins that may be posttranslationally modified (i.e., glycosylation, phosphorylation; refer to detailed discussion in Chapter 5)
3. To compare directly complex protein mixtures (e.g., cellular proteins from cells or tissues from different experimental conditions)
4. To purify proteins directly and preparatively for use in antibody preparation
5. To analyze and determine specificity of antibodies to proteins

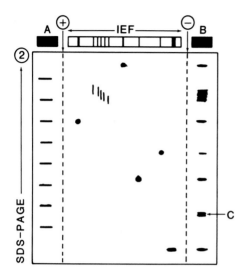

Figure 2.1. Strategy for setting up two-dimensional polyacrylamide gel for maximum information to determine protein sample complexity, homogeneity, or purity. (A) Agarose containing molecular-weight samples. (B) Agarose containing same sample as used in isoelectric focusing (IEF). (C) Protein not visible in two-dimensional pattern. This protein may be too acidic or basic for pH range used in IEF.

V. PEPTIDE MAPPING

While protein electrophoresis methods permit the analysis of non-denatured (native) proteins or denatured reduced proteins that are resolved as individual polypeptides, there is also a frequent need to identify the relatedness of different proteins or to map the functional domains of peptides (Foyt *et al.*, 1985) (see Fig. 2.2). To gather such information, methods have been developed to fractionate and identify specific peptides generated by enzymatic or chemical cleavage of the proteins of interest (see Table 2.1). Such studies have commonly been carried out using both one- and two-dimensional PAGE methods (Ingram, 1956, 1959; Cleveland *et al.*, 1977; Fey *et al.*, 1981, 1984; Bravo *et al.*, 1981). This experimental approach has made possible the identification of homologous peptides between species as well as the detection of specific protein antigens. They have further been used to map functional domains of molecules such as those in myosin light-chain kinase, as shown in Figure 2.2 (Foyt *et al.*, 1985).

This methodology has also been used to map peptides generated by recombinant DNA technology. Frequently, the genes expressed in prokaryotic or eukaryotic expression systems do not contain the entire gene product; therefore, the protein expressed does not have the same molecular weight as that of the native protein. Thus the use of peptide mapping provides important information for determining whether the gene product is from the protein of interest.

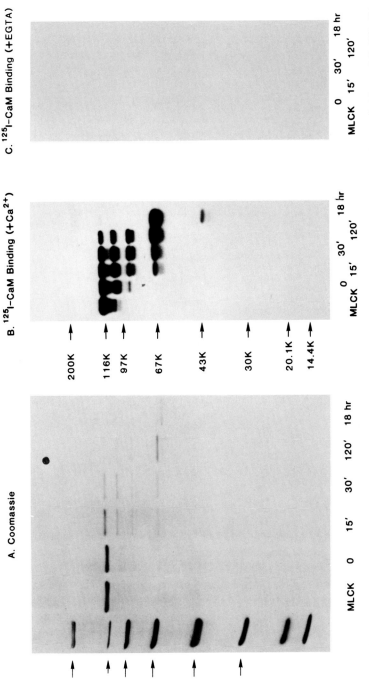

Figure 2.2. Use of peptide mapping to identify functional domains of molecules. V8 protease digests of myosin light-chain kinase (MLCK). (A) Coomassie-stained gel of V8 protease digestion performed in the absence of calmodulin. (B,C) [^{125}I]calmodulin gel overlay of the digests (B, + Ca^{2+}; C, + EGTA). The digest mixtures were subjected to electrophoresis on a 10% Porzio and Pearson sodium dodecyl sulfate–polyacrylamide gel. Methods described in Foyt *et al.* (1985). (Courtesy of H. Foyt and A. R. Means, Department of Cell Biology, Baylor College of Medicine.)

**Table 2.1. Sites of Protein Cleavage by
Chemical or Enzymatic Methods**[a]

Method	NH$_2$ Site–COOH
Chemical	
CNBr	-Met ↓
Formic acid[b]	-Asp ↓ Pro
Enzymatic	
Trypsin[c]	-Lys ↓
Chymotrypsin	-Tyr ↓
	-Phe ↓
	-Trp ↓
Staphylococcus	-Asp ↓
aureus (V8)[d]	-Glu ↓
Papain	Wide specificity

[a]Modified from Fey *et al.* (1984).
[b]For details, refer to Smyth (1967) and Sonderegger
et al. (1982).
[c]The specificity of trypsin can be restricted by blocking amino groups (Butler *et al.*, 1969).
[d]The specificity can be restricted to glutamyl bonds
(Houmard and Drapeau, 1972).

There are now a variety of methods that can be used to carry out
peptide mapping. Earlier studies used paper chromatography but more
recent and popular methods include thin-layer chromatography (TLC)
(Watanabe and Yoshida, 1971), gel electrophoresis (Laemmli, 1970), and
HPLC (Hearn, 1980). The TLC and HPLC methods have been outlined
in detail elsewhere; therefore, only the gel electrophoresis methods are
outlined here.

The method of choice for any given experimental system will depend
on (1) the abundance availability of purified protein, (2) the molecular
weights of the peptides to be analyzed, and (3) the degree of posttranslational modification of a protein that may alter peptide cleavage sites.

A. One-Dimensional Peptide Mapping

While a variety of enzymes described in Table 2.1 can be used to
compare peptide profiles in one-dimensional polyacrylamide gels (see Foyt
et al., 1985), the most widely used method of peptide mapping in polyacrylamide gels has been that described by Cleveland *et al.* (1977). This
method has been popular because enzymatic cleavage of proteins is car-

ried out without prior elution from the polyacrylamide gel. This method is also convenient for use with SDS–PAGE because the *Staphylococcus aureus* V8 protease remains active in the presence of this detergent.

A second chemical method also designed to analyze peptides without the need for elution of proteins from gels was described by Sonderegger *et al.* (1982). This method uses formic acid to cause selective acid hydrolysis of aspartyl–prolyl bonds, which occurs in proteins with an average frequency of 1 per 400 amino acid residues. While this method is simple and gives high reproducibility, it requires long incubations and is only applicable to proteins containing aspartyl–prolyl linkages. Since many proteins do not contain such linkages, this method cannot be universally used.

B. Two-Dimensional Peptide Mapping

Most peptide-mapping methods were originally designed for isolation of polypeptides separated by one-dimensional SDS–PAGE. Although the Cleveland procedure is a rapid and simple method for estimating the relatedness of different proteins, these peptide profiles are not definitive proof of identity. Two-dimensional methods are a more stringent test of polypeptide identity (Hames, 1981). It is now just as practical and frequently advantageous to obtain polypeptides of greater purity if they are isolated from high-resolution two-dimensional polyacrylamide gels. For an even better comparison of peptides from two proteins, the polypeptide can be isolated from two-dimensional polyacrylamide gels and can then be treated enzymatically or chemically to cleave peptides that can be further analyzed by high-resolution two-dimensional PAGE. The gel system may have to be modified, depending on the pI range of peptides.

Additional information can be obtained if peptides are stained with the color-based silver stain, since the color of staining provides yet another criterion for determining the homology of peptides (see discussion in Chapter 4).

C. Limitations of Peptide-Mapping Techniques

The major limitation of this method is that posttranslational modification of proteins such as the presence of carbohydrate side chains may dramatically alter the susceptibility of some proteins to enzymatic or chemical cleavage. This has been shown to be the case with the glycoproteins of the mammalian zona pellucida, the extracellular glycoprotein matrix surrounding the mammalian oocyte (Dunbar *et al.*, 1985).

Distinct as well as related proteins may also have different sensitivities to peptide cleavage reagents. For example, the time of incubation as well as the enzyme concentration used for digestion may alter the peptide pattern obtained by PAGE analysis. It may therefore be necessary to use different enzyme concentrations as well as variable time incubations and different enzymes or chemical reactions to obtain an adequate comparison of the relatedness of different polypeptides. A double-labeling method using two-dimensional PAGE has been adopted as an alternative to peptide mapping for establishing identity between metabolically labeled proteins, especially posttranslationally modified proteins.

In many instances, it may be necessary to carry out amino acid composition and/or sequence studies to determine the true relatedness of proteins.

VI. PURIFICATION OF PROTEINS FROM POLYACRYLAMIDE GELS

Once a protein of interest is identified, it is usually reasonable to purify sufficient quantities for further studies. This can readily be accomplished in most instances by running multiple gels using the multiple gel-casting systems. In order to ensure that a sufficiently pure starting preparation is obtained, it may be necessary to purify the protein partially first. This will make it easier to separate possible contaminating proteins that might migrate in the same region of the two-dimensional PAGE gel and make it difficult to isolate sufficient quantities of purified protein. This can be carried out using conventional protein fractionation methods, such as ammonium sulfate precipitation, separation by differential heat stability, or affinity chromatography. It may also be necessary to adapt gel conditions to separate proteins of interest from others of similar charge and molecular weight. This can be achieved by using a narrow range of ampholytes (in the range of the pI of the protein to be purified) in the IEF dimension and by altering the gradient of acrylamide or percentage of acrylamide in the second dimension.

If a protein is partially purified, it is sometimes possible to use one-dimensional PAGE to purify the protein electrophoretically. (*Note:* The sample should first be analyzed by two-dimensional PAGE to determine that no other contaminating charge protein species of different charges are present.) Larger amounts of protein can be isolated directly by sol-

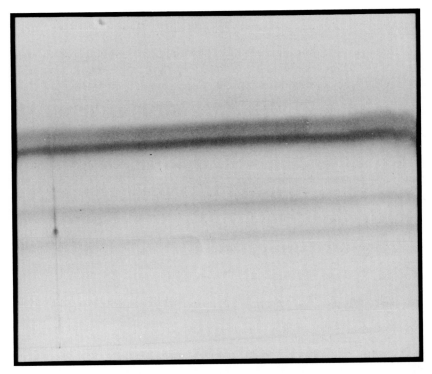

Figure 2.3. **Use of one-dimensional polyacrylamide gel electrophoresis to prepare larger quantities (milligram levels) of protein for electroelution or for immunization.** Sample suspended in agarose and layered across the entire length of the sodium dodecyl sulfate–polyacrylamide gel. Strips of the gel can then be cut out for electroelution (see Appendix 13).

ubilizing the protein sample in SDS (in the presence or absence of a disulfide bond-reducing agent) and heating as for routine SDS–PAGE analysis. This sample can then be spread across the entire surface of a slab polyacrylamide gel (see Fig. 2.3). This method does not require the use of wells in the acrylamide. Following electrophoresis, the entire gel can be stained with Coomassie blue for detection of the protein of interest. Alternatively, if it is desirable not to fix the protein in acetic acid and alcohol, as is required for staining, a portion of the gel can be removed and stained briefly while the remainder of the gel is stored at 4°C to reduce diffusion. The gels can then be matched to determine the location of the protein in the unstained gel.

Once proteins have been isolated by two-dimensional polyacrylamide gels, it is generally necessary to efficiently remove the protein from the polyacrylamide gel matrix for further analyses. (A common exception is the use of proteins for immunization in which it is advantageous to inject the protein suspended in the acrylamide matrix rather than to elute it.) Proteins can be obtained from polyacrylamide gel matrices using a variety of methods, including protein diffusion out of depolymerized or homogenized gels (Hjerten, 1973; Spath and Koblet, 1979; Bridgen, 1976; Goff, 1976; Goerl et al., 1978; Bernabeu et al., 1978; Sreekrishna et al., 1980).

Alternatively, the most efficient method of recovering protein from polyacrylamide gels is the use of electrophoretic elution. A variety of methods and chambers have been developed to recover proteins efficiently (see review by Spiker and Isenberg, 1983). The precise methods that have been used effectively to electroelute larger quantities of protein from many pieces of polyacrylamide gels or for efficient elution of small amounts of proteins for determination amino acid sequence are given in Appendix 13.

VII. OBTAINING PEPTIDES FOR AMINO ACID SEQUENCE

One of the most recent and popular uses for specific peptides obtained from polyacrylamide gels is in obtaining unique amino acid sequences, which can then be employed to prepare oligonucleotide probes. These probes are now commonly used to screen gene libraries for the isolation of genes coding for specific proteins. With the development of technology for the efficient electroelution of proteins and for obtaining amino acid sequences of small quantities of protein (down to 1 pmole) (Hunkapiller et al., 1983), these methods have become increasingly widespread. The equipment and methods for protein elution and preparation for sequencing are outlined in Appendix 13. It is important to appreciate that gel fixation as well as electroelution procedures may alter proteins in a manner that will affect the efficiency of obtaining sequencing of these proteins. Alternate methods for isolation of peptides may therefore be necessary for detailed amino acid analysis (Aebersold et al., 1986).

VIII. PREPARATION AND CHARACTERIZATION OF ANTIBODIES PURIFIED BY POLYACRYLAMIDE GEL ELECTROPHORESIS

One of the most important uses of PAGE, especially high-resolution two-dimensional PAGE, is the preparation of highly purified proteins or peptides for antibody production and characterization. Since it is frequently difficult and time consuming to employ conventional protein-purification methods to purify a protein to homogeneity for use as an immunogen to develop polyclonal or monoclonal antibodies, PAGE provides an excellent option. These methods can be used not only to purify proteins of interest but provide the best method (when used with silver-staining methods) to determine the purity of the protein to be used for immunization. These methods are described in detail in Chapter 13 and Appendix 11.

IX. ANALYSIS OF POSTTRANSLATIONAL MODIFICATION OF PROTEINS

Because many types of posttranslational modification (i.e., glycosylation or phosphorylation) result in changes in the charge of a protein, these can most effectively be analyzed by high-resolution two-dimensional PAGE. Since such modifications can also affect detergent binding properties and will therefore alter the apparent molecular-weight values, a great deal of information can be gained about a protein if these methods are used. The principles and methods that can be applied to analyze these modifications are described in Chapter 5.

X. COMPARISON OF COMPLEX PROTEIN SAMPLES

One of the most effective and popular uses of high-resolution two-dimensional PAGE is for the simultaneous analysis of hundreds of proteins in complex protein mixtures. Such studies have been used to analyze human tissues and fluids as well as animal and plant cell proteins.

Table 2.2 summarizes some selected publications in which two-dimensional PAGE has been used to characterize protein samples directly. The experimental approaches to analyze the synthesis or posttranslational modification of proteins are described in greater detail in Chapter 6. Figures 2.4–2.6 illustrate typical protein patterns obtained from experiments of two-dimensional PAGE.

XI. USE OF MODIFIED ELECTROPHORESIS SYSTEMS TO ANALYZE SPECIFIC PROTEINS

Because proteins have varying physicochemical properties, it is not always possible to use a single electrophoretic method to resolve all pro-

Table 2.2. Examples of Some Applications of Protein Analysis by High-Resolution Two-Dimensional Polyacrylamide Gel Electrophoresis

Application	Topic	Reference
Characterization of human tissue and fluids	Review	N. L. Anderson et al. (1980)
	Review	N. G. Anderson and Anderson (1982)
	Plasma proteins	N. L. Anderson et al. (1984)
	Muscle proteins	Giometti et al. (1983)
	Review	Tracy et al. (1983)
	Cerebrospinal fluid proteins	Goldman et al. (1980)
Identification of abnormal proteins in disease states	Brain proteins	Comings (1981a,b)
	Urine proteins	Edwards et al. (1982)
	Muscle proteins	Giometti et al. (1980)
	Review	Merril and Goldman (1982)
	Review	Tracy and Anderson (1983)
	Serum proteins	Tracy et al. (1982b)
Identification of metabolically labeled proteins	Sea urchin embryogenesis	Bedard and Brandhorst (1983)
	HeLa cell proteins	Bravo and Celis (1982)
	Fibroblast proteins	Dubbelman and Yamada (1982)
	Muscle cell proteins	Garrels (1979)
	Mammalian embryogenesis	Van Blerkom (1978)
	Mammalian stress proteins	Welch et al. (1983)
Identification of phosphorylated proteins	Ovarian proteins	Richards et al. (1983)
	Spermatozoan proteins	Tash et al. (1984)
Identification of glycoproteins	Zona pellucida proteins	Dunbar et al. (1982, 1985a)
	Lymphocyte membrane proteins	Gmunder et al. (1982)
Genetic analysis of human and animal proteins	Brain proteins	Comings (1982a,b)
	Fibroblast proteins	Klose et al. (1983)
	Red blood cell proteins	Knowles et al. (1984)
	Histocompatibility antigens	Osborne et al. (1983)
	Plasma proteins	Pollitt and Bell (1983)
	Plasma proteins	Rosenblum et al. (1983)
Genetic analysis of proteins of viruses and microorganisms	Bacteria proteins	N. G. Anderson (1983); Ayala et al. (1984)
	Neurospora proteins	Nasrallah and Srb (1983)
	Dictyostellium proteins	Prem-Des and Henderson (1983)
	Chlamydomonas proteins	Schmidt et al. (1983)
	Tetrahymena proteins	Suhr-Jessen (1984)
	Ureaplasma proteins	Swenson et al. (1983)
Genetic analysis of plant proteins	Wheat seed proteins	Dunbar et al. (1985a)
Genetic analysis of proteins of cells in tissue culture	Gene amplification of proteins	Cheung et al. (1983)
	Chinese hamster ovary cell proteins	Scoggin et al. (1983)
Analysis of protein changes induced by mutagens or carcinogens	Spermatozoan proteins	Marshall et al. (1983)
Analysis of proteins for forensic science	Human hair proteins	Marshall and Gillespie (1982)

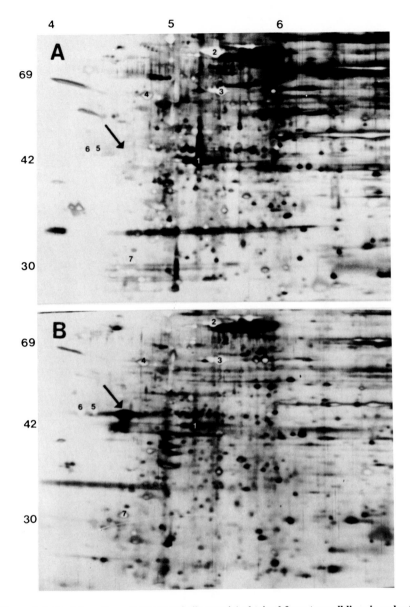

Figure 2.4. Protein patterns (color-based silver stain) obtained from two cell lines in order to examine gene amplification. Two-dimensional gel electrophoretic analysis of the proteins in C1-1D cells (A) as compared with those of the 11AAU/deoxycoformycin-resistant line, B1/3.2 (B). Patterns represent approximately one-third of the proteins resolved by isoelectric focusing in the first dimension, followed by separation in 10–20% gradient polyacrylamide slab gels in the second dimension. Spots 1–7 are included as reference markers, since they are common to both cell types. The arrow indicates the 41,000–43,000-M_r peptide, which is readily apparent in the B1/3.2 cell line, but not in the C1-1D line. (From Yeung *et al.*, 1983.)

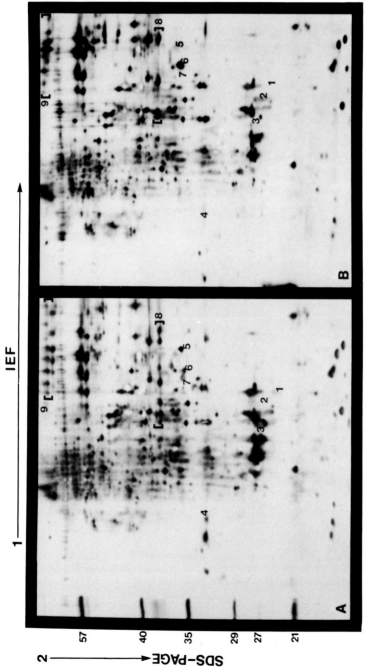

Figure 2.5. Comparison of protein patterns (color-based silver stain) from individual kernels of wheat of two varieties analyzed by high-resolution two-dimensional polyacrylamide gel electrophoresis. The numbers indicate proteins that were found to vary between varieties in all kernels tested. IEF, isoelectric focusing; SDS–PAGE, sodium dodecyl sulfate–polyacrylamide gel electrophoresis. (From Dunbar *et al.*, 1985*b*; VCH Verlagsgesellscaft, mbH.)

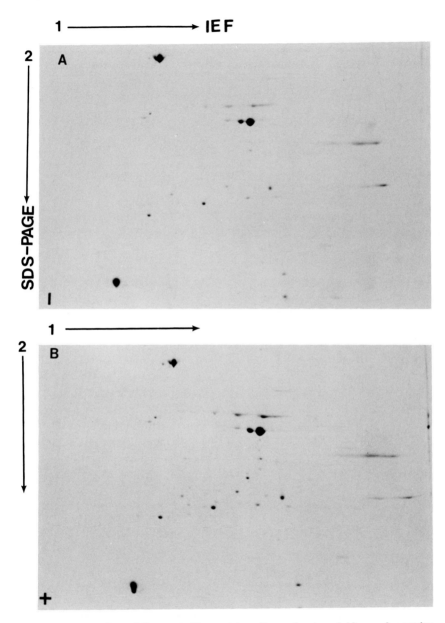

Figure 2.6. Comparison of Coomassie blue protein patterns of water-soluble muscle proteins from (A) control mice and (B) mice exhibiting muscle pathology. (From B. D. Dunbar and D. S. Bundman, unpublished observations.) Although few proteins are observed with this stain, the reproducibility of two-dimensional protein patterns make identification of differences easier.

Table 2.3. Modified Electrophoresis Methods for Separation of Special Proteins and Macromolecules

Proteins separated	Basis of electrophoresis	Reference
Histones	Acid–urea gels	Panyim and Chalkley (1969); Hardison and Chalkley (1978)
	Triton–acid–urea gels	Zweidler (1978)
Low-molecular-weight peptides	Modified acrylamide buffer systems	Porzio and Pearson (1977); DeWald *et al.* (1986)
Lipopolysaccharides	Modified sodium dodecyl sulfate	Hitchcock (1983)

teins optimally from a given sample. Different investigators have therefore designed different electrophoretic systems to analyze specific proteins of interest, some of which are outlined in Table 2.3. One such system is for the analysis of histones. Histones are highly basic proteins modified by such reactions as acetylation or phosphorylation. Because these proteins tend to be insoluble in the absence of denaturing reagents, they are frequently analyzed by SDS–PAGE. Because the use of detergents may mask the presence of charge variants, methods have been developed that permit size and charge fractionation while preventing protein aggregation (see discussion by Hames, 1981).

Other protein fractionation methods have shown that the presence of impurities in electrophoresis reagents can sometimes aid in the separation of proteins that are otherwise difficult to separate (such as α- and β-tubulin). While many of these methods may provide important information about some proteins of interest, there may frequently be a great deal of laboratory-to-laboratory variability associated with some of these methods.

XII. CLINICAL APPLICATION OF TWO-DIMENSIONAL POLYACRYLAMIDE GEL ELECTROPHORESIS

Only recently has two-dimensional PAGE been applied to the detailed analysis of clinical samples (see Table 3.2). Initial analyses have elucidated the complexities of protein mixtures that are present. These protein patterns are further complicated if sensitive silver stains are used. While this technology is being adopted by increasing numbers of laboratories (see reviews by Anderson and Anderson, 1982*a,b;* Tracy and Anderson, 1983; Tracy *et al.*, 1983*a,b*), its use in clinical laboratory testing is currently limited for the following reasons:

1. *Lack of standardization between laboratories:* Although there is now a strong move for standardization of methodology among laboratories, there are still many variations in the precise methodologies used by different laboratories. The development of equipment for large-scale electrophoresis and for improved internal standardization for both charge and molecular weight has improved comparisons of gel patterns among laboratories. It is anticipated that such standardization will greatly improve the interpretation of results obtained by two-dimensional PAGE analysis of clinical samples.

2. *Technical difficulty in carrying out analyses:* As there are many variables in two-dimensional PAGE procedures, the methodology requires technical personnel well versed in the basic principles of this technology in order to troubleshoot as well as to interpret results adequately. Investigators must be willing to troubleshoot if optimal results are not obtained.

It is one of the goals of this text to help investigators and their personnel understand this technology so that initial setup time will be minimal and quality of data obtained optimal.

It is anticipated that as the number of laboratories using standardized methods increases, the data base required for basic as well as for clinical research will be adequate to take full advantage of these technologies. One of the most promising aspects of high-resolution two-dimensional PAGE is that it can be used to purify proteins directly as well as prepare and characterize specific antibodies to them. These antibodies can then be used to set up larger-scale clinical assays that require less time, expense, and expertise.

REFERENCES

Aebersold, R. H., Teplow, D. B., Hood, L. E., and Kent, S. B. H., 1986, Electroblotting onto activated glass. High efficiency preparation of proteins from analytical sodium dodecyl sulfate polyacrylamide gels for direct sequence analysis, *J. Biol. Chem.* **261:**4229–4238.

Anderson, N. G., 1983, High-resolution protein separation and identification methods applicable to virology, in: *Current Topics in Microbiology and Immunology,* Vol. 104 (M. Cooper, P. H. Hofschneider, H. Koprowski, F. Melchers, R. Roh, H. G. Schwerger, P. K. Vogt, and R. Zinhernagel, eds.), pp. 197–217, Springer-Verlag, Berlin.

Anderson, N. G., and Anderson, N. L., 1982a, The human protein index, *Clin. Chem.* **28:**739–748.

Anderson, N. G., and Anderson, N. L., 1982b, The human protein index and the molecular pathology data base, *Med. Lab.* **11:**75–94.

Anderson, N. L., Edwards, J. J., Giometti, C. S., Willard, K. E., Tollaksen, S. L., Nance, S. L., Hickman, B. J., Taylor, J., Coulter, B., Scandora, A., and Anderson, N. E., 1980, High-resolution two-dimensional electrophoretic mapping of human proteins, in: *Electrophoresis '79* (B. Radola, ed.), pp. 313–328, W. deGruyter, Berlin.

Anderson, N. L., Tracy, R. P., and Anderson, N. G., 1984, High-resolution two-dimensional electrophoretic mapping of plasma proteins, in: *The Plasma Proteins*, 2nd ed., Vol. 4 (F. W. Putnam, ed.), pp. 221–270, Academic Press, New York.

Ayala, J. A., dePedro, M. A., and Vasquez, D., 1984, Application of a charge/size two-dimensional gel electrophoresis system to the analysis of the penicillin-binding proteins of *Escherichia coli*, *FEBS Lett.* **168**:93–96.

Bedard, P. A., and Brandhorst, B. P., 1983, Patterns of protein synthesis and metabolism during sea urchin embryogenesis, *Dev. Biol.* **96**:74–83.

Bernabeu, C., Conde, F. P., and Vazquez, D., 1978, Extraction of pure ribosomal protein and removal of Coomassie Blue from acrylamide gels, *Anal. Biochem.* **84**:97–102.

Bravo, R., and Celis, J. E., 1982, Updated catalogue of HeLa cell proteins: Percentages and characteristics of the major cell polypeptides labeled with a mixture of 16 ^{14}C-labeled amino acids, *Clin. Chem.* **28**:766–781.

Bravo, R., Fey, S., Small, J. V., Mose Larsen, P., and Celis, J. E., 1981, Coexistence of three major isoactins in a single Sarcoma 180 cell, *Cell* **25**:195–202.

Bridgen, J., 1976, High sensitivity amino acid sequence determination. Application to proteins eluted from polyacrylamide gels, *Biochemistry* **15**:3600–3604.

Butler, P. J. G., Harris, J. I., Hartley, B. S., and Leberman, R., 1969, The use of maleic anhydride for the reversible blocking of amino groups in polypeptide chains, *Biochem. J.* **112**:679–689.

Cheung, C., Ingolia, D. E., Bobonis, C., Dunbar, B. S., Riser, M. E., Siciliano, M. J., and Kellums, R. E., 1983, Selective production of adenosine deaminase in cultured mouse cells, *J. Biol. Chem.* **258**:8338–8345.

Cleveland, D. W., Fischer, S. G., Kirschner, M. W., and Laemmli, U. K., 1977, Peptide mapping by limited proteolysis in sodium dodecyl sulfate and analysis by gel electrophoresis, *J. Biol. Chem.* **252**:1102–1106.

Comings, D. E., 1982a, Two-dimensional gel electrophoresis of human brain proteins. I. Technique and nomenclature of proteins, *Clin. Chem.* **28**:782–789.

Comings, D. E., 1982b, Two-dimensional gel electrophoresis of human brain proteins. III. Genetic and non-genetic variations in 145 brains, *Clin. Chem.* **28**:798–804.

DeWald, D. B., Adams, L. D., and Pearson, J. D., 1986, A non-urea electrophoretic gel system for resolution of polypeptides of $M_r = 2,000$ to $M_r = 200,000$, *Anal. Biochem.* **154**:502–508.

Dubbelman, T. M., and Yamada, K. M., 1982, A survey of differences between membrane polypeptides of transformed and non-transformed chick embryo fibroblasts, *Biochim. Biophys. Acta* **692**:177–187.

Dunbar, B. S., Liu, C., and Sammons, D. W., 1981, Identification of the three major proteins of porcine and rabbit zonae pellucidae by high-resolution two-dimensional gel electrophoresis: Comparison with serum, follicular fluid, and ovarian cell proteins, *Biol. Reprod.* **24**:1111–1124.

Dunbar, B. S., Dudkiewicz, A. B., and Bundman, D. S., 1985a, Proteolysis of specific porcine zona pellucida glycoproteins in boar acrosin, *Biol. Reprod.* **32**:619–630.

Dunbar, B. D., Bundman, D. S., and Dunbar, B. S., 1985b, Identification of cultivar specific proteins of winter wheat (*T. aestivum*, L.) by high-resolution two-dimensional polyacrylamide gel electrophoresis and color-based silver stain, *Electrophoresis* **6**:39–43.

Edwards, J. J., Anderson, N. G., Tollaksen, S. L., Van Eschenbach, A. E., and Guevara,

J., Jr., 1982, Proteins of human urine. II. Identification by two-dimensional electrophoresis of a new candidate marker for prostatic cancer, *Clin. Chem.* **28**:160–163.

Fey, S. J., Bravo, R., Mose Larsen, P., Bellatin, J., and Celis, J. E., 1981, (^{35}S)Methionine-labeled polypeptides from secondary mouse kidney fibroblasts: Coordinates and one-dimensional peptide maps of some major polypeptides, *Cell Biol. Int. Rep.* **5**:491–500.

Fey, S. J., Bravo, R., Mose Larsen, P., and Celis, J. E., 1984, Correlation between mouse and human two-dimensional gel patterns: Peptide mapping of proteins extracted from two-dimensional gels, in: *Two-Dimensional Gel Electrophoresis of Gel Proteins: Methods and Applications* (J. E. Celis and R. Bravo, eds.), pp. 170–189, Academic Press, New York.

Foyt, H., Guerriero, V., and Means, A. R., 1985, Functional domains of chicken gizzard myosin light chain kinase, *J. Biol. Chem.* **260**:7765–7774.

Garrels, J. I., 1979, Changes in protein synthesis during myogenesis in a clonal cell line, *Dev. Biol.* **73**:134–152.

Giometti, C. S., Barany, M., Danon, M. J., and Anderson, N. G., 1980, Muscle protein analysis. II. Two-dimensional electrophoresis of normal and diseased human skeletal muscle, *Clin. Chem.* **26**:1152–1155.

Giometti, C. S., Danon, M. J., and Anderson, N. G., 1983, Human muscle proteins: Analysis by two-dimensional electrophoresis, *Neurology (New York)* **33**:1152–1156.

Gmunder, H., Lerch, P., and Lesslauer, W., 1982, Human lymphocyte membrane proteins treated with neuraminidase, *Biochim. Biophys. Acta* **693**:359–363.

Goerl, M., Welfe, H., and Bielka, H., 1978, Preparative two-dimensional polyacrylamide gel electrophoresis of rat liver ribosomal proteins and determination of their amino acid composition, *Biochim. Biophys. Acta* **519**:418–427.

Goff, C. G., 1976, Histones of *Neurospora crassa, J. Biol. Chem.* **251**:4131–4138.

Goldman, D., Merril, C. R., and Ebert, M. H., 1980, Two-dimensional gel electrophoresis of cerebrospinal fluid proteins, *Clin. Chem.* **26**:1317–1322.

Hames, B. D., 1981, Peptide mapping using SDS–PAGE, in: *Gel Electrophoresis of Proteins: A Practical Approach* (B. D. Hames and D. Rickwood, eds.), pp. 219–228, IRL Press, Oxford, England.

Hardison, R., and Chalkley, R., 1978, Polyacrylamide gel electrophoretic fractionation of histones, *Methods Cell Biol.* **17**:235–251.

Hearn, M. J. W., 1980, The use of reversed phase high performance liquid chromatography for the structural mapping of polypeptides and proteins, *J. Liq. Chromatogr.* **3**:1255–1276.

Hedrick, J. L., and Smith, A. J., 1968, Size and charge isomer separation and estimation of molecular weights of proteins by disc gel electrophoresis, *Arch. Biochem. Biophys.* **126**:155–164.

Hitchcock, P. J., 1983, Aberrant migration of lipopolysaccharide in sodium dodecyl sulfate polyacrylamide gel electrophoresis, *Eur. J. Biochem.* **133**:685–688.

Hjerten, S., 1973, Preparative gel electrophoresis: Recovery of proteins, in: *Methodological Developments in Biochemistry,* Vol. 2 (E. Reid, ed.), pp. 39–47, Academic Press, New York.

Houmard, J., and Drapeau, G. R., 1972, Staphylococcal protease: A proteolytic enzyme specific for glutamyl bonds, *Proc. Natl. Acad. Sci. USA* **69**:3506–3509.

Hunkapiller, M W., Lujan, E., Ostrander, F., and Hood, L. E., 1983, Isolation of microgram quantities of proteins from polyacrylamide gels for amino acid sequence analysis, *Methods Enzymol.* **91**:227–236.

Ingram, V. M., 1956, A specific chemical difference between the globins of normal human and sickle cell anaemia haemoglobin, *Nature (Lond.)* **178**:792–794.

Ingram, V. V., 1959, Abnormal human haemoglobin. III. The chemical difference between normal and sickle cell haemoglobins, *Biochim. Biophys. Acta* **36**:402–411.

Klose, J., Willers, I., Singh, S., and Goedde, H. W., 1983, Two-dimensional electrophoresis of soluble and structure bound proteins from cultured human fibroblasts and hair root cells: Qualitative and quantitative variation, *Hum. Genet.* **63**(3):262–267.

Knowles, W. J., Bologna, M. L., Chasis, J. A., Marchesi, S. L., and Marches, V. T., 1984, Common structural polymorphisms in human erythrocyte spectrin, *J. Clin. Invest.* **73**:973–979.

Laemmli, U. K., 1970, Cleavage of structural proteins during the assembly of the head of bacteriophage T4, *Nature (Lond.)* **277**:680–685.

Marshall, R. C., and Gillespie, J. M., 1982, Comparison of samples of human hair by two-dimensional electrophoresis, *J. Forensic Sci. Soc.* **22**:377–385.

Marshall, R. R., Raj, A. S., Grant, F. J., and Heddle, J. A., 1983, The use of two-dimensional electrophoresis to detect mutations induced in mouse spermatogonia by ethylnitrosourea, *Can. J. Genet. Cytol.* **25**(5):457–466.

Merril, C. R., and Goldman, D., 1982, Quantitative two-dimensional protein electrophoresis for studies of inborn errors of metabolism, *Clin. Chem.* **28**(4):1015–1020.

Nasrallah, J. B., and Srb, A. M., 1983, Two-dimensional electrophoresis of plasma membranes showing differences among wild-type and abnormal ascopore mutant strains of *Neurospora crassa*, *J. Bacteriol.* **155**(3):1393–1398.

Osborne, B. A., Lunnej, J. K., Pennington, J., Sachs, D. H., and Rudikoff, S., 1983, Two-dimensional gel analysis of swine histocompatibility antigens, *J. Immunol.* **131**(6):2939–2944.

Panyim, S., and Chalkley, R., 1969, High-resolution acrylamide gel electrophoresis of histones, *Arch. Biochem. Biophys.* **130**:337–346.

Pollitt, C. C., and Bell, K., 1983, Characterization of the alpha-1-protease inhibitor system in thoroughbred horse plasma by horizontal two-dimensional (ISO-DALT) electrophoresis. I. Protein staining, *Anim. Blood Groups Biochem. Genet.* **14**(2):83–105.

Porzio, M.A., and Pearson, A.M., 1977, Improved resolution of myofibrillar proteins with sodium dodecyl sulfate polyacrylamide gel electrophoresis, *Biochim. Biophys. Acta* **490**:27–34.

Prem-Das, O., and Henderson, E. J., 1983, Developmental regulation of Dictyostellium plasma membrane proteins, *J. Cell Biol.* **97**(5):1544–1558.

Richards, J. S., Seghal, N., and Tash, J. S., 1983, Changes in content and cAMP-dependent phosphorylation of specific proteins in granulosa cells of preantral and preovulatory ovarian follicles and corpora lutea, *J. Biol. Chem.* **258**:5227–5232.

Rosenblum, B. B., Neel, J. V., and Hanosh, S. M., 1983, Two-dimensional electrophoresis of plasma polypeptides reveals "high" heterogeneity indices, *Proc. Natl. Acad. Sci. USA* **80**(16):5002–5006.

Schmidt, R. J., Richardson, C. B., Gillham, N. W., and Boynton, J. E., 1983, Sites of synthesis of chloroplast ribosomal proteins in *Chlamydomonas*, *J. Cell Biol.* **96**:1451–1463.

Scoggin, A. U., Paul, S., Miller, Y. E., and Patterson, D., 1983, Two-dimensional electrophoresis of peptides from human CHO cell hybrids containing human chromosome 21. *Somat. Cell Genet.* **9**(6):687–697.

Smyth, D. G., 1967, Techniques in enzymic hydrolysis, *Methods Enzymol.* **11**:214–231.

Sonderegger, P., Jaussi, R., Gehring, H., Brunschweiler, K., and Christen, P., 1982, Peptide mapping of protein bands from polyacrylamide gel electrophoresis by chemical cleavage in gel pieces and re-electrophoresis, *Anal. Biochem.* **122**:298–301.

Spath, P. J., and Koblet, H., 1979, Properties of SDS–polyacrylamide gels highly crosslinked with N,N-diallytartardiamide and the rapid isolation of macromolecules from the gel matrix, *Anal. Biochem.* **93**:275–285.

Spiker, S., and Isenberg, I., 1983, Preparative polyacrylamide gel electrophoresis, *Methods Enzymol.* **91:**214–216.

Sreekrishna, K., Jones, C. E., Guetzow, K. A., Prasad, M. R., and Joshi, V. C., 1980, A modified method for amino acid analysis with ninhydrin of Coomassie-stained proteins from polyacrylamide gels, *Anal. Biochem.* **103:**55–57.

Suhr-Jessen, P. B., 1984, Stage specific changes in protein synthesis during conjugation in *Tetrahymena thermophilia, Exp. Cell Res.* **151**(2):374–383.

Swenson, C. E., Van Hamont, J., and Dunbar, B. S., 1983, Specific protein determined by two-dimensional gel electrophoresis and differences among strains of *Ureaplasma urealyticum* as sensitive silver stain, *Int. J. Syst. Bacteriol.* **33:**417–421.

Tash, J. S., Kakar, S. S., and Means, A. R., 1984, Flagellar motility requires the cAMP-dependent phosphorylation of a heat stable NP-40 soluble 56 Kd protein, akokinin, *Cell* **38:**551–559.

Tracy, R. P., and Anderson, N. L., 1983, Applications of two-dimensional gel electrophoresis in the clinical laboratory, in: *Clinical Laboratory Annual* (H. A. Homburger and J. G. Butsaks, eds.), pp. 101–130, Appleton-Century-Crofts, East Norwalk, Connecticut.

Tracy, R. P., Curie, R. M., and Young, D. S. (1982*a*), Two-dimensional electrophoresis of serum specimens from a normal population, *Clin. Chem.* **28**(4):890–899.

Tracy, R. P., Currie, R. M., Kyle, R. A., and Young, D. S. (1982*b*), Two-dimensional gel electrophoresis of serum specimens from patients with monoclonal gammopathies, *Clin. Chem.* **28**(4):900–907.

Tracy, R. P., Currie, R. M., Hill, H. D., and Young, D. S., 1982*c,* Methods of two-dimensional gel electrophoresis in the clinical laboratory, in: *Protides of Biological Fluids,* Vol. 30 (H. Peeters, ed.), pp. 581–586, Pergamon Press, Oxford.

Tracy, R. P., Currie, R. M., Hill, H. D., and Young, D. S., 1983, Two-dimensional gel electrophoresis in clinical chemistry, in: *Fifth Colloquium of International Biologie Prospective,* pp. 79–84.

Van Blerkom, J. V., 1978, Methods for the high-resolution analysis of protein synthesis: Applications to studies of early mammalian development, in: *Methods in Mammalian Reproduction* (J. C. Daniel, Jr., ed.), pp. 67–109, Academic Press, New York.

Watanabe, S., and Yoshida, A., 1971, Microscale peptide mapping method for identification of variant proteins, *Biochem. Genet.* **5:**541–547.

Welch, W. J., Garrels, J. I., Thomas, G. P., Lin, J. J., and Feramisco, J. R., 1983, Biochemical characterization of mammalian stress proteins and identification of two stress proteins as glucose and Ca^+ ionophore-regulated proteins, *J. Blol. Chem.* **258:**7102–7111.

Willard, K. E., 1982, Two-dimensional analysis of human lymphocyte proteins. III. Preliminary report on a marker for the early detection and diagnosis of infectious mononucleosis, *Clin. Chem.* **28:**1031–1035.

Yeung, C., Ingola, D., Bobonis, C., Dunbar, B., Riser, M., Siciliano, M., and Kellems, R., 1983, Selective overproduction of adenosine deaminase in cultured mouse cells, *J. Biol. Chem.* **258**(13):8338–8345.

Zweidler, A., 1978, Resolution of histones by polyacrylamide gel electrophoresis in presence of nonionic detergents, in: *Methods in Cell Biology,* Vol. 17 (G. Stein, J. Stein, and L. J. Kleinsmith, eds.), pp. 223–233, Academic Press, New York.

Chapter 3

Sample Preparation for Electrophoresis

I. QUANTITATION OF PROTEIN IN SAMPLES

While standardized protein assays are used in nearly every laboratory, the limitations of many of these assays are frequently overlooked. It is important that an investigator realize that most protein assays are relative and are based on the measurement of particular amino acid residues and/or peptide bonds. The estimation of protein in different assays will therefore depend on the abundance of those amino acids in the protein sample. Because different protein assays may detect different amino acid residues, totally different estimations of protein content may be obtained for the same sample when different methodologies are used. It may therefore be necessary to use more than one protein assay to determine which is optimal for your needs. Table 3.1 summarizes amino acid residues or other chemical bases for protein estimation by different protein assay methods. The commonly used methods for determining total protein have been described elsewhere by Lowry *et al.* (1951), Warburg and Christian (1941), and Bradford (1976) and have recently been reviewed in detail by Petersen (1983). This latter review also presents a detailed list of chemicals including buffers and detergents that interfere with some of these protein assays.

It has been recommended that the Folin phenol quantitation method and its modifications be used for general use in order to promote consistency in protein measurements between laboratories (Petersen, 1983). This recommendation has been made because this is the method that has been most frequently used and that is most widely applicable and consistent with different proteins.

If an investigator is interested in a rapid method for protein detection and/or a quantitative estimate of protein content, the Coomassie Blue-based protein assay described by Bradford (1976) is frequently a method of choice. Reagents for this assay are commercially available from BioRad Laboratories. Furthermore, this method has now been modified so that

Table 3.1. Summary of Protein-Determination Methods

Method	Amino acid residue measured	References
Folin phenol	Aromatic amino acids: (tyrosine and tryptophan) (slower reaction with copper chelates of peptide chain and/or polar side chains)	Lowry et al. (1951); Peterson (1977, 1979, 1983)
UV absorption ($A_{280/260}$ nm) ratio	A_{280}: Tyrosine and tryptophan A_{260}: Nucleic acid interference measurement	Warburg and Christian (1941); Peterson (1983)
UV absorption (224–236 nm)	A_{240}: Tyrosine, tryptophan, phenylalanine, histidine, methionine, cysteine, and peptide bonds A_{230}: Nucleic acid absorption (minimum)	Groves et al. (1968); Petersen (1983)
UV absorption (280/205 nm)	A_{280}: Tyrosine and tryptophan A_{205}: Peptide bond (subject to interference by nucleic acids)	Scopes (1974); Petersen (1983)
Coomassie Blue		Bradford (1976); Read and Northcote (1981)
O-Phthaldehyde fluorescent	NH_2-terminal amino acids and ϵ-amino group of lysine	Butcher and Lowry (1976); Robrish et al. (1978)

it can be used directly to measure protein levels in the presence of the solubilization buffers used for both isoelectric focusing (IEF) and sodium dodecyl sulfate–polyacrylamide gel electrophoresis (SDS–PAGE) (Ramagli and Rodriguez, 1985).

II. PRINCIPLES OF PROTEIN SOLUBILIZATION

While most investigators pay a great deal of attention to the methodology to be used in protein fractionation, frequently less consideration is given to sample preparation. We have found invariably that this is the most important consideration for obtaining optimal results with one-dimensional as well as IEF dimension for two-dimensional PAGE.

The choice of method to be used for solubilization is dependent on the physicochemical nature of the proteins to be analyzed. There is not a universal method that will result in optimal solubilization and/or resolution of all protein samples analyzed by electrophoretic methods. It is therefore essential that initial studies be carried out to determine which procedure is best for a given protein sample. Occasionally, it may be necessary to sacrifice some resolution in order to obtain adequate solubilization. The following guidelines should be helpful in understanding the principles of solubilization that are critical in carrying out most electrophoretic studies.

Most methods used for protein solubilization require the denaturation of proteins. In living organisms, however, proteins exist in their native states which reflect the location and the physiological conditions for normal functioning of that protein. It must therefore be accepted that the properties of any protein or polypeptide that is removed from its physiological environment for analysis *in vitro* may be quite different from its properties *in vivo*.

In order to understand the basis of protein denaturation, it is first necessary to appreciate the basic terminology that is commonly applied to the "native" states of proteins (Lapanje, 1978).

1. *Primary structure:* Denotes the sequence of amino acid residues in the protein molecule
2. *Secondary structure:* Signifies that hydrogen bonds are involved in a higher-order structure, such as α-helix formation
3. *Tertiary structure:* Refers to the pattern of folding into a compact, globular molecule
4. *Quaternary structure:* Signifies the noncovalent association of two or more subordinate entities, subunits, which may or may not be identical

The process of molecular denaturation is a complex one involving differing degrees of alterations of these structures. Many aspects of these have been previously outlined (Tanford, 1968, 1970; Lapanje, 1978). While predictions for the sites of action of protein denaturants have been obtained from data using model compounds, these usually cannot be applied directly to complex macromolecules.

Denaturation has been defined as the reversible or irreversible process in which the native protein conformation undergoes a major change without rupturing the primary covalent bonds, i.e., without altering the primary structure (Lapanje, 1978). This definition does not, however, consider the existence of disulfide bonds and their role in maintenance of the protein conformation, which are also critical in most studies of protein structure, as these may crosslink two different parts of the same polypeptide chain or two different chains. In order to unfold the polypeptide chain, it is therefore necessary to reduce the disulfide bonds. This is generally carried out using reagents such as β-mercaptoethanol or di-thiothreitol (DTT).

III. EXPERIMENTAL SOLUBILIZATION OF PROTEIN SAMPLES

Experimentally, denaturation can be accomplished in many ways, depending on the complexity of the macromolecule. Common methods have included thermal denaturation, pH, urea, guanidinium chloride, inorganic salts, organic solvents, and ionic as well as nonionic detergents. Frequently, combinations of these methods are used to improve the solubilization of complex protein mixtures for optimal separation. Only those methods commonly used for electrophoretic analysis of proteins are discussed here.

A. Denaturation by Urea

Urea ($H_2NCO\ NH_2$) has been and remains one of the most widely used denaturants. Extensive physicochemical studies have been carried out to analyze the extent of denaturation of a variety of different proteins (see review by Lapanje, 1978). The extent of protein denaturation in urea has been shown to depend on temperature, pH, and ionic strength. Usually, its effectiveness is obtained only at relatively high concentrations (≥ 8 M), and there still remain many proteins that do not undergo complete denaturation (Tanford, 1968). As with other denaturants, most physicochemical studies on the denaturation efficacy of urea have been carried out using highly purified homogeneous proteins.

Urea is most commonly used to solubilize protein for IEF. Urea works by disrupting hydrogen bonds and has the advantage for some applications that it does not affect the intrinsic charge of proteins so that separation of constituent polypeptides will be on the basis of size and charge in contrast to the use of SDS (Hames, 1981). The major disadvantage is that urea must be present during electrophoresis to maintain the denatured state of the protein.

B. Denaturation by Guanidinium Chloride

Guanidinium chloride (GndCl) is one of the most effective denaturants. In general, most proteins will become randomly coiled in 6 M GndCl.

Because this salt interferes with most electrophoresis procedures, it is not commonly used in the methods outlined in this text.

C. Denaturation by Detergents

Ionic as well as nonionic detergents have long been known to bring about various degrees of denaturation. In general, however, the ionic detergents are more active. In addition to bringing about denaturation, it has been demonstrated that, at low concentrations, they can protect proteins from heat coagulation or from denaturation by urea and GndCl (see reviews by Lapanje, 1978; Tanford, 1968, 1970; Steinhart and Reynolds 1969).

1. Nonionic Detergents

Nonionic detergents are commonly used to solubilize proteins from membranes or to solubilize proteins in a manner that may not affect the biological activity of that protein. To date, the most commonly used detergents include Triton X-100 (Makino et al., 1973; Dewald et al., 1974), Lubrol and Sarkosyl (cited in Newby and Chrambach, 1979), and Nonidet P-40.

While the solubilization of complex proteins is not fully understood, some basic studies using purified proteins have been carried out to help determine the mechanisms by which these detergents interact with proteins. Triton X-100 has been concluded to interact predominantly by binding hydrophobically to certain proteins (Green, 1971). In these studies, it was observed that some proteins, including horse myoglobin and human IgG, did not bind Triton-X, while other proteins did bind (in decreasing order: human sera albumin, bovine serum albumin, and β-lactoglobulin).

Equilibrium dialysis studies that compared the binding of the anionic detergent, deoxycholate, with the nonionic detergent, Triton X-100, to bovine serum albumin (BSA) have demonstrated that there are four principal ligand sites for each detergent and about 14 weaker sites for deoxycholate (Makino et al., 1973). It was further found that cooperative binding of ligand with unfolding did not occur with these two detergents, presumably because these two detergents have a relatively low critical micelle concentration. Therefore, a monomeric concentration sufficient for the cooperative mode of binding cannot be attained. This offers an advantage, however, when using deoxycholate and Triton X-100 for the solubilization of membrane proteins.

2. Anionic Detergents

In general, ionic detergents are more effective in solubilizing proteins than are nonionic detergents. The most commonly used detergents are sodium dodecyl (lauryl) sulfate (NaDodSO$_4$; CH$_3$(CH$_2$)$_{10}$CH$_2$OSO$_3$Na; SDS) and sodium deoxycholate. Although sodium deoxycholate is an ionic detergent, it resembles nonionic detergents in its low denaturing effects. Since sodium deoxycholate also does not bind to membrane proteins in significant quantities, it also causes fewer conformational changes than strongly ionic detergents such as SDS. This detergent has been shown to dissociate larger proteins into subunits, unfolding the polypeptide chains to form rodlike complexes of proteins to which SDS molecules are bound, mainly by hydrophobic bonds (Reynolds and Tanford, 1970a,b).

To date, several detailed investigations have used the interaction between BSA and anionic detergents as a model to understand the mechanism of denaturation (Decker and Foster, 1966; Reynolds et al., 1967; see also review by Tanford, 1968). In its native state, BSA has been reported to contain a small number (10–11) of binding sites with large binding constants for single detergent ions. These binding sites are destroyed if the native structure is lost (Decker and Foster, 1966), and as a result, native BSA is stabilized against denaturation by urea, pH, and so forth, when very low concentrations of anionic detergents are present (Lovrien, 1963; Markus et al., 1964).

However, with an excess of detergent, there are altered conformations that possess the ability to bind a much larger number of detergent ions and for which binding properties are much different from those of the native protein. The binding of detergent ions to the native protein is thought to occur one molecule at a time, whereas the binding to the denatured protein is highly cooperative. The existence of discrete protein–detergent complexes further suggests that the association between detergent ions is also important when the degree of binding to the molecule is small. A higher than expected ratio of binding of detergent molecules to a single binding site has implicated the formation of *detergent micelles* at this site (see review by Tanford, 1968).

Other studies have shown that the binding of SDS to protein (β-lactoglobulin) yields similar behavior to that of proteins bound by *n*-octylbenzenesulfonate (Hill and Briggs, 1956). At low concentrations, such a protein molecule was found to interact with two or three molecules of detergent and then to undergo a change resulting in an increased ability to bind detergent. At the same time, a third type of binding was identified described as "micellar" in nature involving the interaction between detergent molecules already bound to the protein and those in solution. This micellar binding was shown to be reversible with a limit at the critical micelle concentration (CMC).

More recently, detailed studies using equilibrium dialysis have been carried out to elucidate the role of SDS binding to proteins with special reference to PAGE (Takagi *et al.*, 1975). In these studies, binding of detergent (in 50 mM sodium phosphate buffer at 25°C, pH 7) was found to occur in two phases with an initial gradual increase in binding (0.3–0.6 g/g protein) followed by a subsequent steep increase up to 1.2–1.5 g/g protein. The binding that varied from protein to protein was found to be complete at an SDS concentration below the CMC.

In general, it is apparent that the process of protein denaturation and solubilization is an extremely complex one and that no generalizations can be made that will be optimal for every protein sample, especially for complex protein mixtures.

D. Solubilization of Membrane Proteins

The solubilization of membrane proteins for purification is accompanied by several difficulties. Some proteins are associated with the external or internal surface of the membrane while others are intercalated in the lipid bilayer. The strategy for isolating your protein of interest will depend on its interaction with the lipid bilayer of the cellular membranes. Since proteins may be associated with membranes in a variety of ways, it will be necessary to design your experiments to dissociate or solubilize your protein accordingly. (A detailed list of membrane proteins and the methods for isolation has been provided by Guidotti, 1972.) A minor portion of membrane proteins are loosely bound to, or associated with, membranes. Some of these can be dissociated by changing the ionic strength or pH of the medium or by chelating the divalent cations involved in the binding of some proteins to the membrane (see review by Razin, 1972). For example, basic proteins such as cytochrome C are positively charged at neutral pH and are therefore bound to membrane phospholipids by electrostatic bonds that can be disrupted by high salt concentrations.

Although PAGE is commonly used to separate and identify soluble enzymes, membrane enzymes and other proteins are more difficult to identify, presumably due to their lipoprotein character. Because large lipoprotein molecules or aggregates are unlikely to enter even 5–10% acrylamide gels or fail to resolve adequately if they do, it is necessary to solubilize or disaggregate these prior to electrophoresis (Dewald *et al.*, 1984). Many solubilization methods used to extract proteins from the lipid portion of the membranes are frequently accompanied by irreversible inactivation of membrane enzymes or alteration in the physicochemical properties of the molecules to be analyzed. Essentially, all methods using organic solvents for lipid extraction cause some degree of protein denaturation. Organic solvents (such as 2-chloroethanol or *n*-butanol) have been used to solubilize membranes because they have the capacity to separate membrane lipids from proteins (Zahler and Weibel, 1970). Acetic acid has also been used, since at low concentrations (10%) it has been reported to act as an acid by dissociating the loosely bound membrane proteins. At high concentrations (>50%), however, it acts mainly as an organic solvent and weakens the hydrophobic bonds within the protein subunits as well as between the protein and lipid molecules (see discussion in Razin, 1972).

Generally, ionic detergents are more effective than nonionic detergents in solubilizing membranes. If maintenance of enzymatic or biological activity is important, however, ionic detergents may be unacceptable. As with other protein samples, the degree of membrane solubilization depends on the detergent concentration. This optimal concentration has been shown to be dependent on the weight ratio of detergent to membrane protein rather than on the absolute detergent concentration (Selinger *et al.*, 1969; Engelman *et al.*, 1967). Membranes have also been solubilized with a combination of detergents (e.g., deoxycholate and Triton-X). Recently, a nondenaturing zwitterionic detergent, 3-[(3-chloramidopropyl)dimethylammonio]-1-propane sulfonate (CHAPS), has been reported to be effective at breaking protein–protein interactions (Hjelmeland, 1980). While this detergent may not be universally useful, it behaves as a zwitterion and possesses no net charge over a large pH range. It therefore does not interfere with ion-exchange or IEF methods to the extent that many other detergents do.

In general, we have found that either the urea solubilization or SDS solubilization methods described in Appendix II for IEF or for SDS–PAGE are adequate for optimal resolution of most membrane proteins when one- or two-dimensional PAGE is used.

IV. FACTORS CRITICAL FOR SAMPLE SOLUBILIZATION PRIOR TO ISOELECTRIC FOCUSING

Although many methods have been developed to solubilize proteins and protein samples effectively, some of these may cause artifacts or interfere with protein migration during IEF. It is frequently necessary to use multiple sample solubilization procedures to determine whether you are getting maximal scientific information from your analyses.

Such striking examples of solubilization variability prior to two-dimensional PAGE are shown in Figures 3.1 and 3.2. In these studies, phosphorylation of membrane proteins was analyzed by high-resolution two-dimensional PAGE. The resolution by silver-staining patterns of the urea-solubilized proteins was lower than that of the SDS-solubilized proteins (data not shown). However, the major phosphoprotein was not adequately solubilized when the urea procedure was used. Likewise, two abundant zona pellucida glycoproteins were not solubilized by the urea method as compared with the SDS method and did not enter the IEF gel. While these situations do not always occur, it is important that a new investigator in this area be aware of the possible limitations of using a single solubilization method.

The apparent isoelectric point (pI) of proteins may also vary depending on the method used for solubilizing the protein sample. Because SDS is an anionic (negatively charged) detergent that binds to proteins, these SDS-coated proteins may initially move very rapidly toward the cathode in the electrophoretic field until the SDS is removed from the protein. At the point that the detergent is removed, the charge on the protein may change. If a protein is basic, it must migrate back to its relative pH range. Since detergent binding varies from protein to protein and the relative molecular weights of proteins may vary as well, different proteins may take markedly different times to focus and to reach a steady state in the IEF gel. It may therefore be necessary to vary electrofocusing times when first starting to optimize protein-separation methods. *These variables illustrate the importance of the use of internal IEF charge standards.* Charge standards will greatly aid in the interpretation of protein patterns and will help you evaluate the possibility that detergents or other ions in your samples may adversely effect your isoelectric focusing (see Fig. 3.3).

V. ARTIFACTS OF PROTEIN SOLUBILIZATION

A. Detergent Concentration and Interaction

The solubilization of proteins with detergents is a complex process (see Section II). Inadequate levels of detergent may result in inadequate solubilization, so that proteins do not enter the polyacrylamide gel. If more than one detergent is used (e.g., anionic and nonionic), they may

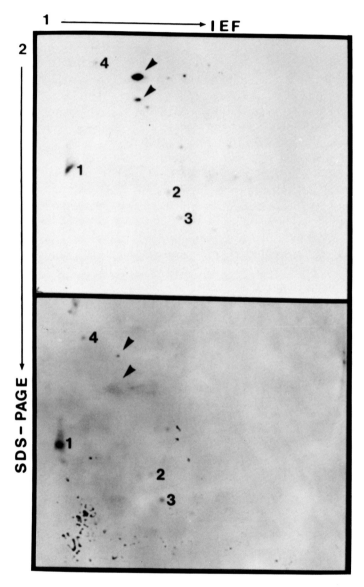

Figure 3.1. Autoradiographs of high-resolution two-dimensional polyacrylamide gel electrophoresis (PAGE) of phosphorylated proteins from dog sperm membranes following the sodium
dodecyl sulfate (SDS) solubilization (A) as compared with the urea solubilization (B) method
for isoelectric focusing (IEF)(Appendix 1). Note that the major phosphoproteins (arrows) are
not solubilized as efficiently by the urea method as by the SDS method. (Methods for
phosphorylation described in Tash *et al.*, 1985.) (Autoradiographs courtesy of Dr. J. Tash,
Department of Cell Biology, Baylor College of Medicine.)

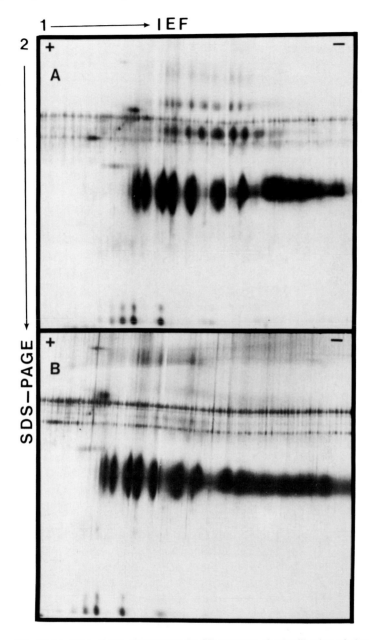

Figure 3.2. Comparison of two-dimensional PAGE patterns of solubilization of chemically deglycosylated porcine zona pellucida proteins following (A) SDS solubilization (Appendix 1) and (B) urea solubilization (Appendix 1). Note that two major families of proteins are not solubilized by the urea method. IEF, isoelectric focusing; PAGE, polyacrylamide gel electrophoresis; SDS, sodium dodecyl sulfate.

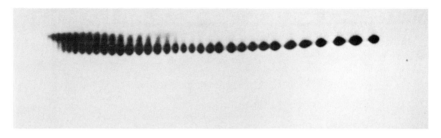

Figure 3.3. Use of internal standards to compare protein samples during isoelectric focusing.
The internal charge standards are generated by heating a basic protein (creatine kinase) in
urea for different periods of time. The charge on the protein molecules will be altered as a
result of carbamylation. A pool of differentially carbamylated protein species is added to
the sample to be analyzed by isoelectric focusing (IEF) using Pharmalyte, pH 3–10 ampholyte (Pharmacia Fine Chemical Co.). Because a single protein species is used, the charge
train can easily be identified and used to compare charge distributions of protein in different
gels. (From Tollaksen *et al.*, 1981; Verlag Chemie GmbH, Weinheim, Germany.)

also interact if used simultaneously and form aggregates that will interfere
with proteins entering polyacrylamide gels or will cause streaking.

B. Chemical Modification of Proteins

Many side chains of amino acid residues of globular proteins easily
react with a variety of specific chemical reagents. While these reactions
can readily be used to detect and quantitate levels of proteins (see Chapter
6), they can also induce changes in protein that may alter charge or
structure. These changes would therefore affect the mobilities of proteins
in IEF or in SDS–PAGE. The most common modifications that can occur
during protein sample preparation or solubilization are carbamylation and
deamidation.

Carbamylation is the process of adding a carbamyl group (NH_2CO^-)
to an amino group. This can occur when protein samples are allowed to
heat in the presence of urea. Under these conditions, the cyanate generated from urea will cause carbamylation of proteins and may dramatically alter their charge properties. For these reasons, proteins should
never be heated in the presence of urea prior to analysis by IEF unless
you are processing proteins to make internal charge standards (see Appendix 3).

Deamidation is the removal of an amino group (NH_2) from an amino
acid. Deamidation occurs when primary amino groups are converted into

hydroxyl groups. This can occur if proteins are treated by acetic acid or nitric acid in dilute HCl (Haurowitz and Tunca, 1945).

C. "Insoluble" Proteins

Some proteins are difficult to solubilize or may interact with other proteins even in the presence of high concentrations of either detergents or denaturing reagents, or both. It is important that a systematic study be carried out utilizing different solubilization methods in conjunction with one- as well as two-dimensional PAGE to determine the best method with which to analyze your protein samples. In extreme cases, it may be necessary to carry out proteolysis of samples in order to at least evaluate the peptide composition of proteins that cannot be evaluated with conventional methods.

VI. SUBFRACTIONATION OF SAMPLES FOR ELECTROPHORESIS

It is critical that the investigator who is just initiating experiments be aware of the limitations of the analysis of total cellular or biological fluid samples that may contain thousands of proteins. Most of the abundant cellular proteins are "housekeeping" proteins that are essential for the life of the cell. It is frequently not possible to observe discrete minor differences which might be brought about by experimental variables if total cellular protein is analyzed. It is therefore usually necessary to carry out some form of protein subfractionation prior to protein analysis in order to analyze minor proteins in a sample.

A. Organelle Fractionation Using Centrifugation and Electrophoresis

Classic methods for subfractionating cellular components utilizing centrifugation have been described in detail previously (see reviews by Fleischer and Kervina, 1974; Reid and Williamson, 1974). This method is summarized in Figure 3.4. While these methods have been invaluable in studies on the structure and function of specific organelles of cells, they are time consuming and limited for several aspects of protein analysis: (1) only small numbers of samples can be easily processed; (2) these methods take long periods of time and therefore endogenous proteolysis can cause alterations in protein prior to electrophoretic analysis; and (3)

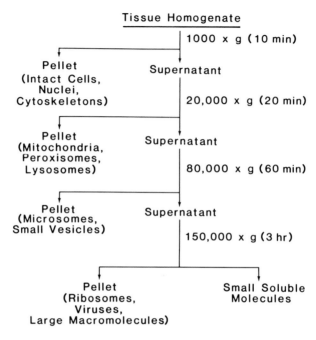

Figure 3.4. Diagram of classical centrifugation method for subfractionating cellular components.

detailed analyses must be carried out to prove the quality of the preparation (e.g., electron microscopy and enzyme identification).

B. Water-Soluble Fractionation

For the rapid fractionation of cells or tissue, it is possible to homogenize rapidly (before or after pulverization of tissue in liquid nitrogen) in distilled H_2O. The samples can then be centrifuged at 100,000–200,000 × g for 1 hr, and the water-insoluble as well as the water-soluble fractions can be analyzed. Figure 3.5 illustrates muscle proteins that have been fractioned in this manner. Note that the protein patterns of the total muscle sample are markedly different from those of the water-soluble and

Figure 3.5. Identification of muscle proteins by two-dimensional gel electrophoresis and the color-based silver stain. (A) Water-insoluble muscle proteins. (B) Water-soluble fraction obtained from muscle proteins. A, actin; T, tubulin; P, parvalbumin.

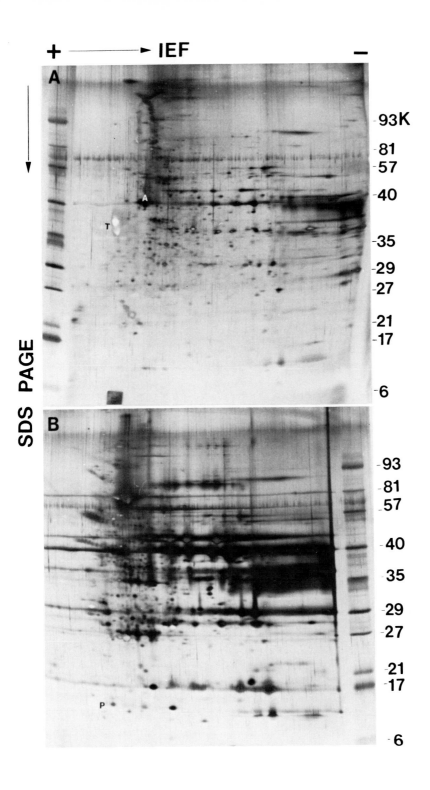

-insoluble fraction. Because this method is rapid and simple, it is usually the method of choice if tissue or cell samples are available only in limited amounts.

Alternative protein fractionation methods have been developed for protein analysis that rely on independent and more expedient chemical extractions (Lenstra and Bloemendal, 1983). These methods have permitted the fractionation of cells into subcellular fractions corresponding to water-soluble, membrane, microfilament, intermediate filament, microtubule, polysomal, and nucleic protein fractions. Care must be taken throughout these procedures to ensure that proteins are not proteolyzed or chemically modified prior to protein analyses. More recently, new methods and equipment have been developed for the electrophoretic purification of cells and membranes (Menashi and Crawford, 1980; Menashi et al., 1981; Morre and Heidrich, 1983). These new methods of free-flow electrophoresis should prove invaluable in future studies of cell isolation and fractionation.

C. Chemical Extraction Methods for Protein Analysis

A variety of methods are available for extracting proteins directly from cells or tissues for analysis by PAGE. Generally, these require multiple samples for different extraction procedures. One series of chemical extractions has been developed to analyze seven subcellular fractions corresponding to (1) water-soluble proteins, (2) membrane proteins, (3) microfilaments and other deoxycholate-soluble proteins, (4) intermediate filaments, (5) microtubules, (6) polysomes, and (7) nuclei (Lenstra and Bloemendal, 1983).

It should be pointed out that while cell fractionation methods are important for analysis and comparison of complex protein samples, it is more difficult to establish definitively that a protein is a constituent of a particular cellular organelle. It will eventually be necessary to prove the cellular localization of a protein using such methods as immunocytochemical localization.

D. Differential Absorption of Proteins to Charged Matrices

It is frequently advantageous to fractionate a concentrated protein partially before it is analyzed by electrophoretic methods. Optimal methods involve procedures that are rapid and that take a minimal amount of time to carry out, so that large numbers of samples can be processed without proteolysis of proteins, and so forth. Affigel blue (BioRad) is frequently used to remove albumin from serum or plasma samples to facilitate identification of minor serum proteins (Tracy and Young, 1984). We have found, however, that other proteins may also be removed by this or other charged bead matrices commonly used for protein fractionation.

A convenient, inexpensive method for rapidly concentrating proteins for analysis by electrophoresis using Celite (diatomaceous earth) was recently developed (Maresh and Dunbar, 1986). This method is useful because small amounts of protein may be absorbed and concentrated with minimal loss of protein and without the simultaneous concentration of salts that will interfere with isoelectric focusing. This method is outlined in Appendix 2.

REFERENCES

Bradford, M. M., 1976, A rapid and sensitive method for the quantitation of microgram quantities of protein utilizing the principle of protein–dye binding, *Anal. Biochem.* **72:**248–254.

Butcher, E. C., and Lowry, O. H., 1976, Measurement of nanogram quantities of protein by hydrolysis followed by reaction with orthophthalaldehyde or determination of glutamate, *Anal. Biochem.* **76:**502–523.

Decker, R. V., and Foster, J. F., 1966, The interaction of bovine plasma albumin with detergent anions. Stoichiometry and mechanism of binding of alkylbenzenesulfonates, *Biochemistry* **5:**1242–1254.

Dewald, B., Dulaney, J. T., and Touster, O., 1974, Solubilization and polyacrylamide gel electrophoresis of membrane enzymes with detergents, *Methods Enzymol.* **32:**82–91.

Engelman, D. M., Terry, T. M., and Morowitz, H. J., 1967, Characterization of the plasma membrane of mycoplasma. I. Sodium dodecyl sulfate solubilization, *Biochim. Biophys. Acta* **135:**381–390.

Fleischer, S., and Kervina, M., 1974, Subcellular fractionation of rat liver, *Methods Enzymol.* **31:**6–41.

Green, F. A., 1971, Interactions of a non-ionic detergent II with soluble proteins, *J. Colloid. Interface Sci.* **35:**481–485.

Groves, W. E., Davis, F. C., Jr., and Sells, B. H., 1968, Spectrophotometric determination of microgram quantities of protein without nucleic acid interference, *Anal. Biochem.* **22:**195–210.

Guidotti, G., 1972, Membrane proteins, *Annu. Rev. Biochem.* **41:**731–752.

Hames, B. D., 1981, An introduction to polyacrylamide gel electrophoresis, in: *Gel Electrophoresis of Proteins: A Practical Approach* (B. D. Hames and D. Rickwood, eds.), pp. 1–91, IRL Press, Oxford, England.

Haurowitz, F., and Tunca, M., 1945, The linkage of glutamine in proteins, *Biochem. J.* **39:**443–445.

Hill, R. M., and Briggs, D. R., 1956, A study of the interaction of n-Octylbenzene-p-sulfonate with β-lactoglobulin, *J. Am. Chem. Soc.* **78:**1590–1597.

Hjelmeland, L. M., 1980, A non-denaturing zwitterionic detergent for membrane biochemistry: Design and synthesis, *Proc. Natl. Acad. Sci. USA* **77**(1)**:**6368–6370.

Lapanje, S., 1978, *Physicochemical Aspects of Protein Denaturation*, John Wiley & Sons, New York.

Layne, E., 1957, Spectrophotometric and turbidimetric methods for measuring proteins, *Methods Enzymol.* **3**:447–454.

Lenstra, J. A., and Bloemendal, H., 1983, Topography of the total protein population from cultured cells upon fractionation by chemical extractions, *Eur. J. Biochem.* **135**:413–423.

Lovrien, R., 1963, Interaction of dodecyl sulfate anions of low concentration with alkaline bovine serum albumin, *J. Am. Chem. Soc.* **85**:3677–3682.

Lowry, O. H., Rosebrough, N. J., Farr, A. L., and Randall, R. J., 1951, Protein measurement with the folin phenol reagent, *J. Biol. Chem.* **193**:265–275.

Makino, S., Reynolds, J. A., and Tanford, C., 1973, The binding of deoxycholate and Triton X-100 to proteins, *J. Biol. Chem.* **248**:4926–4932.

Maresh, G., and Dunbar, B. S., 1986, A simple method for the concentration and analysis of secreting proteins from tissue cell cultures (submitted for publication).

Markus, G., Love, R. L., and Wissler, F. C., 1964, Mechanism of protection by anionic detergents against denaturation of serum albumin, *J. Biol. Chem.* **239**:3687–3693.

Menashi, S., and Crawford, N., 1980, Isolation of platelet surface and intracellular membranes by high voltage free-flow electrophoresis: Differential distribution of cytoskeletal components, *Eur. J. Cell Biol.* **22**:598.

Menashi, S., Weintroub, H., and Crawford, N., 1981, Characterization of human platelet surface and intracellular membranes isolated by free-flow electrophoresis, *J. Biol. Chem.* **256**(8):4095–4101.

Morre, D. J., Morre, D. M., and Heidrich, H. G., 1983, Subfractionation of the rat liver golgi apparatus by free-flow electrophoresis, *Eur. J. Cell Biol.* **31**:263–274.

Newby, A. C., and Chrambach, A., 1979, Disaggregation of adenylate cyclase during polyacrylamide gel electrophoresis in mixtures of ionic and non-ionic detergents, *Biochem. J.* **177**:623–630.

Peterson, G. L., 1977, A simplification of the protein assay method of Lowry et al. which is more generally applicable, *Anal. Biochem.* **83**:346–356.

Peterson, G. L., 1979, Review of the folin phenol protein quantitation method of Lowry, Rosebrough, Farr and Randall, *Anal. Biochem.* **100**:201–220.

Peterson, G. L., 1983, Determination of total protein, *Methods Enzymol.* **91**:95–119.

Ramagli, L. S., and Rodriguez, L. V., 1985, Quantitation of microgram amounts of protein in two-dimensional polyacrylamide gel electrophoresis sample buffer, *Electrophoresis* **6**:559–563.

Razin, S., 1972, Reconstitution of biological membranes, *Biochim. Biophys. Acta* **265**:241–296.

Read, S. M., and Northcote, D. H., 1981, Minimization of variation in the response to different proteins of the Coomassie blue G dye-binding assay for protein, *Anal. Biochem.* **116**:53–64.

Reid, E., and Williamson, R., 1974, Centrifugation, *Methods Enzymol.* **31**:713–733.

Reynolds, J. A., and Tanford, C., 1970a, Binding of dodecyl sulfate to proteins at high binding ratios. Possible implications for the state of proteins in biological membranes, *Proc. Natl. Acad. Sci. USA* **66**:1002–1007.

Reynolds, J. A., and Tanford, C., 1970b, The gross conformation of protein–sodium dodecyl sulfate complexes, *J. Biol. Chem.* **245**:5161–5165.

Reynolds, J. A., Herbert, S., Polet, H., and Steinhart, J., 1967, The binding of divers detergent anions to bovine serum albumin. *Biochem.* **6**:930–947.

Robrish, S. A., Kemp, C., and Bouen, W. H., 1978, The use of the o-phthalaldehyde reaction as a sensitive assay for protein and to determine protein in bacterial cells and dental plaque, *Anal. Biochem.* **84**:196–204.

Scopes, R. K., 1974, Measurement of protein by spectrophotometry at 205 nm, *Anal. Biochem.* **59**:277–282.

Selinger, Z., Klein, M., and Amsterdam, A., 1969, Properties of particles prepared from sarcoplasmic reticulum by deoxycholate, *Biochem. Biophys. Acta* **183**:19–26.

Steinhart, J., and Reynolds, J. A., 1969, *Multiple Equilibrium in Proteins,* Academic Press, New York.

Takagi, T., Tsujii, K., and Shirahama, K., 1975, Binding isotherms of sodium dodecyl sulfate to protein polypeptides with special reference to SDS–polyacrylamide gel electrophoresis, *J. Biochem. (Tokyo)* **77**:939–947.

Tanford, C., 1968, Protein denaturation, *Adv. Protein Chem.* **23**:121–282.

Tanford, C., 1970, Protein denaturation, *Adv. Protein Chem.* **24**:2–97.

Tash, J. S., Kakar, S. S., and Means, A. R., 1984, Flagellar motility requires the cAMP-dependent phosphorylation of a heat stable NP-40 soluble 56 Kd protein, axokinin, *Cell* **38**:551–559.

Tollaksen, S. L., Edwards, J. J., and Anderson, N. G., 1981, The use of carbamylated charge standards for testing batches of ampholytes used in two-dimensional electrophoresis, *Electrophoresis* **2**:155–160.

Tracy, R. P., and Young, D. S., 1984, Clinical applications of two-dimensional gel electrophoresis, in: *Two-Dimensional Gel Electrophoresis of Proteins: Methods and Applications* (J. E. Celis and R. Bravo, eds.), pp. 193–240, Academic Press, New York.

Warburg, O., and Christian, W., 1941, Isolierung und Kristallisation des Garungsferments Enolase, *Biochem. Z.* **310**:384–421.

Zahler, P., and Weibel, E. R., 1970, Reconstitution of membranes by recombining proteins and lipids derived from erythrocyte stroma, *Biochem. Biophys. Acta* **219**:320–338.

Chapter 4

Protein Detection in Polyacrylamide Gel Electrophoresis

I. INTRODUCTION

An investigator must take several factors into account for the process of staining and recording the location of bands after electrophoresis (Wilson, 1979, 1983). These should include (1) the selection and availability of an appropriate dye of satisfactory quality that will give adequate sensitivity; (2) the selection of a solvent that will provide a suitable dye concentration, solubility, and stability; (3) the formation of a stable dye–protein complex; (4) the easy removal of excess free dye for easy detection of stained proteins; and (5) storage of gels for further reference or for permanent recording by photography or analysis by computer, densitometry, or spectrophotometry.

The method of choice of protein detection will depend to a great extent on the experimental question being asked. For example, if one is interested in analyzing changes in single or limited number of abundant proteins, a less sensitive protein stain, such as Coomassie Blue, is generally adequate. However, if one is interested in assessing the purity of a protein to be used as an immunogen for antibody preparation, the most sensitive stains, such as silver staining, should be used. If an investigator is interested in synthesis of specific proteins in response to experimental or environmental factors, isotopic labeling is generally the method of choice. It should be stressed that, as with protein-assay methods, protein-staining methods do not stain all proteins uniformly; therefore, a direct quantitation of proteins in gels is usually only a relative measurement and is dependent on the amount of staining reagent bound to a protein. The advantages and disadvantages of each method are described in this chapter.

II. COOMASSIE BLUE STAINING METHODS

Until recently, when the more sensitive silver staining methods were developed, Coomassie Blue staining methods were most commonly used to detect proteins in polyacrylamide gel electrophoresis. These methods were recently summarized by Wilson (1979, 1983).

Coomassie Blue R and Coomassie Blue G are biochemically distinct dyes commonly used under the tradenames of Coomassie Brilliant Blue R250 and G250. Coomassie is the registered trademark for a number of acid wool dyes held by the Imperial Chemical Industries (ICI), which no longer makes these dyes. The number 250 is the ICI dye strength indicator, which is frequently misused for dyes of other strengths. These dyes have also been called a variety of other names, including Xylene brilliant cyanine GB or G, Serva Blau (Blue) R or G, PAGE blue 83 or 90, Supran-

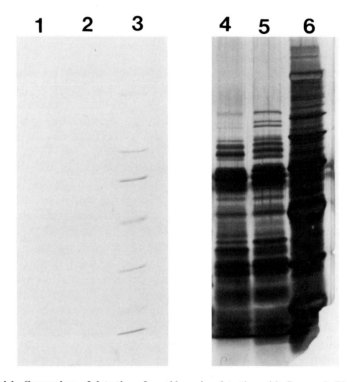

Figure 4.1. Comparison of detection of peptides using detection with Coomassie Blue (lanes 1–3) followed by staining of the identical gel with the color-based silver stain method (lanes 4–6). Note the dramatic differences in the sensitivity of staining of proteins with the silver stain method.

ocyanin GB or G, Kenacid blue R, brilliant blue R, and Microme #1137 or 1137. The color index number for these dyes is CI42660, acid blue 83 for Coomassie Blue R, and CI42655 acid blue 90 for Coomassie Blue G.

Because of the complexities of the nomenclature of these dyes as well as the quality and purity of commercially available reagents, the investigator should take care to evaluate the quality and structure of the dyes to be used for protein staining.

The protein staining of protein will occur through the ionic interaction between the basic amino acids and the acid dyes as well as secondary interactions due to hydrogen bonding, vander Waals attraction, and hydrophobic bonding between dyes and protein as well as between free dye and dye already associated with the protein. Because these interactions may be affected differently by pH, ionic concentration, solvents, and other factors, it is likely that most proteins will bind different amounts of dyes under different conditions (Wilson, 1983). In general, 0.2–0.5 μg protein in a sharp band or spot can be detected when the dyes and staining are quantitated for some proteins for up to 15 μg (Hames, 1981).

Staining variations can also occur when sodium dodecyl sulfate (SDS) is bound to proteins. This is particularly a problem if poor-quality SDS is used that contains impurities that may affect both the relative mobility of the protein as well as the intensity of the stain (Matheka *et al.*, 1977). A variety of fixation and staining methods may be necessary if this is a problem (see review by Wilson, 1983).

In summary, while Coomassie Blue R is commonly used and produces staining bands, and these methods are relatively straightforward, this dye does not have ideal properties for quantitative staining of proteins.

III. SILVER STAINING OF PROTEINS IN POLYACRYLAMIDE GELS

A. Introduction

The Coomassie Blue staining methods were the most popular direct methods of protein detection in polyacrylamide gels until more sensitive silver staining methods were reported. Reports that modified histological silver stain methods have a 100-fold increased sensitivity over Coomassie Blue methods (Merril *et al.*, 1979; Switzer *et al.*, 1979) led to an explosion of variations of silver stain methods (see reviews by Merril and Goldman, 1984; Sammons *et al.*, 1984). Figures 4.1 and 4.2 compare a protein sample stained with Coomassie Blue and the color-based silver stain (Gelcode).

Figure 4.6. Use of color-based silver stain to analyze related proteins in two-dimensional polyacrylamide gel electrophoresis. Water-soluble proteins of mouse muscle were separated by two-dimensional polyacrylamide gel electrophoresis. Proteins exhibiting charge heterogeneity due to such posttranslational modifications as phosphorylation or glycosylation (brackets) stain the same color. If proteins are overloaded (arrows) the center of the spot may stain a different color than the edge (halo effect).

Currently, there are more than 100 variations of silver staining methods, making a detailed review of this subject beyond the scope of this text. These stains have, for the most part, been developed from photographic chemistry in which silver crystals are used to form images. Because there are so many variations in silver staining methods, it is important that the investigator take into account such factors as polyacrylamide gel thickness and gel pore size as well as the sensitivity and resolution of the silver stain method to be employed.

These numerous variables associated with these staining methods have frustrated many investigators, in particular, those just beginning to use these techniques. Recently, however, methods for some of these

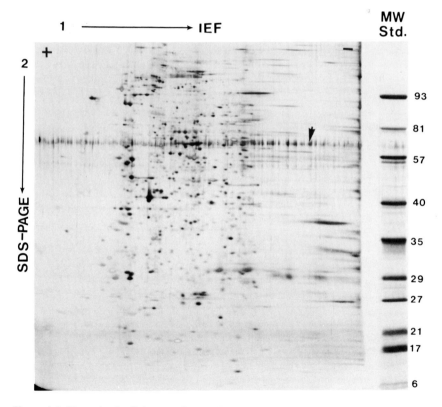

Figure 4.4. Example of cellular protein sample analyzed by high-resolution two-dimensional polyacrylamide gel and the color-based silver stain. The molecular markers, which are color coded (available from Health Products, Inc.), are prepared in agarose and are run in one dimension at the side of the isoelectric focusing (IEF) gel during the second-dimension run. Arrow indicates a mercaptan artifact.

staining procedures have been systematically developed so that they are more reproducible and easier to use.

After considerable effort in trying many of these silver stains, investigators in my laboratory have settled on two methods that are the most straightforward and reproducible: (1) the monochromatic silver stain developed by Wray *et al.* (1981), and (2) the polychromatic stain described by Sammons *et al.* (1981), known as Gelcode (Health Products, Inc., and Pierce Chemical Co.). These methods are outlined in detail in Appendix 9.

A monochromatic stain may be desired if complex one-dimensional polyacrylamide gels are used to analyze proteins. The polychromatic stain, however, is not only more sensitive than the other original stains (Ochs *et al.*, 1981), but provide a third dimension (color) in protein resolution, since proteins stain five basic colors—black, blue, brown, red, and yellow (see Fig. 4.3). Furthermore, the posttranslational modification, such as glycosylation or phosphorylation, of proteins does not appear to alter the color of protein staining; therefore, charge forms of protein molecules can be identified in two-dimensional polyacrylamide gels (Dunbar *et al.*, 1985; see discussion in Chapter 5). Finally, this method does not involve a destaining step as most conventional silver stain methods, a process that can account for a great deal of variability between experiments as well as among different laboratories. Color-coded molecular-weight markers are also available from Health Products, Inc., for silver stain methods. Unlike most commercially available molecular-weight markers, these proteins migrate as distinct as opposed to diffuse bands (Fig. 4.4). Finally, the color-based silver stain makes it possible to detect minor proteins in overloaded gels. Because of the contrast of colors, it is possible to detect many proteins that would be indistinguishable by monochromatic staining techniques (see Figs. 4.5 and 4.6).

B. Factors Affecting Silver Stain Reactions

Although the precise mechanisms for all silver stain procedures that give them their unique staining properties are unknown, the sensitivity and color-staining properties are dependent on the fixatives and reagent quality used as well as the manner in which the silver is allowed to interact with the protein (Merril and Goldman, 1984; Sammons *et al.*, 1984). In some methods the silver reacts primarily with the protein at the surface of the gel. This is why dirt or fingerprints are detected on the surface of gels if extreme care is not taken in handling gels. In the color-based method described by Sammons *et al.* (1984), however, the silver nitrate is allowed

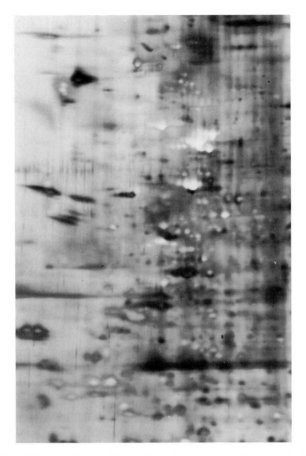

Figure 4.5. Use of the color-based silver stain to detect proteins in an overloaded gel. Most of the proteins identified in this portion of a two-dimensional gel would be indistinguishable if a monochromatic stain were used, but the contrast given by color differentiation permits easy detection of many minor proteins.

to permeate into the acrylamide matrix for a longer period of time before it is fixed to the protein. Because color is dependent on light reflections, it is the intercalation of the silver into the protein molecule that will affect the color exhibited by that protein. (Because the excess silver is washed from these gels, there are also fewer problems with interference from surface dirt or fingerprints.)

For some stains, especially those using ammonium hydroxide reagents, it is critical that solutions be made fresh just prior to staining if optimal results are to be obtained. The quality of water as well as agitation during staining are also factors that are most critical for success of these procedures. Because water quality can vary dramatically in different regions of the country, it may be necessary to try different water-purification systems before optimal results are achieved. Initially, it is advisable to use commercial reagents and silver stain molecular-weight standards, such as those available from Health Products, Inc. (Fig. 4.4), to establish your techniques until you have tested your water quality and technique.

C. Molecules and Artifacts Stained by Silver Stain Methods

In addition to the detection of proteins in PAGE, silver stain methods have also been used to detect nucleic acids (Merril and Goldman, 1984) as well as lipopolysaccharides (Hitchcock and Brown, 1983). In protein samples, the presence of nucleic acids can (as demonstrated in Chapter 10, Fig. 10.8) make it difficult to identify proteins clearly. We have observed, however, that the color-based silver stain, Gelcode (Appendix 7), stains nucleic acids a distinct dark color, so that proteins can be easily detected.

In addition to these molecules, we have observed that this stain will bind to a variety of unknown molecules that can be from impure reagents or organics that may come from plastics or glues used in electrophoresis chambers.

Common staining artifacts are those that are found at the region of the polyacrylamide gel corresponding to molecular weights of approximately 68,000 and 54,000 (Merril, 1982; Ochs, 1983; Marshall and Williams, 1984). This artifact appears to be associated with the use of mercaptoethanol in the isoelectric focusing (IEF) equilibration buffer and can be reduced or eliminated if mercaptoethanol is omitted from this buffer. (Note, however, that the omission of this reducing agent may result in irregular protein patterns due to disulfide bond reformation. It will thus be necessary to test your protein sample to determine whether this is a problem.) This artifact has been attributed to contamination by keratins from skin if gloves are not used throughout all electrophoretic procedures (Ochs, 1983). This observation was further documented by the use of antibodies to keratins. We have observed, however, that many control nonimmune sera will also detect this artifact (B. S. Dunbar and D. S. Bundman, unpublished observations). More detailed studies appear to be necessary before these contaminating molecules which stain with sensitive silver stain techniques are clearly defined. It is apparent that the use of two-dimensional PAGE makes it easier to determine which are artifacts of staining and which are specific proteins of interest.

IV. MISCELLANEOUS PROTEIN-STAINING METHODS

A large number of protein-staining methods have been reported for use with PAGE. Many of these methods have been summarized by Wilson (1983) and Hames (1981). Because most of these protein-staining methods are insensitive as compared with Coomassie Blue and even more so as compared with the more recently developed silver stains, they are not discussed in detail in this text.

Amido black (acid Black 1) and Fast Green FCF (Food green 3) are two dyes that have been used in the past but are used more infrequently now. The precise methods for these staining procedures are given by Wilson (1983). Fluorescent dye methods have also been used to detect proteins. These methods include the labeling of proteins before electrophoresis (Ragland et al., 1974; Schetters and McLeod, 1979; Tjissen and Kurstak, 1979) as well as following electrophoresis (Hartman and Udenfriend, 1969; Jackowski and Liew, 1980; Carson, 1977). The amido black stain can also be used to stain protein transfers on nitrocellulose.

A number of direct protein-detection methods have also been described that use the precipitation of SDS polypeptide complexes with K^+ ions (Nelles and Bamburg, 1976), sodium acetate (Higgins and Dahmus, 1979), or cationic surfactant (Takagi et al., 1977). The advantages of these methods are that fixation procedures using acids and alcohols are not needed prior to protein detection. The disadvantages are the lack of sensitivity as well as limited resolution of protein detection.

REFERENCES

Carson, S. D., 1977, Hydrazinocridine staining of proteins and glycoproteins in polyacrylamide gels, *Anal. Biochem.* **78:**428–435.

Dunbar, B. S., Dudkiewicz, A. B., and Bundman, D. S., 1985, Proteolysis of specific porcine zona pellucida glycoproteins by boar acrosin, *Biol. Reprod.* **32:**619–630.

Hames, B. D., 1981, An introduction to polyacrylamide gel electrophoresis, in: *Gel Electrophoresis of Proteins: A Practical Approach* (B. D. Hames and D. Rickwood, eds.), pp. 1–91, IRL Press, Oxford, England.

Hartman, B. K., and Udenfriend, S., 1969, A method for immediate visualization of proteins in acrylamide gels and its use for preparation of antibodies to enzymes, *Anal. Biochem.* **30:**391–394.

Higgins, R. C., and Dahmus, M. E., 1979, Rapid visualization of protein bands in preparative SDS–polyacrylamide gels, *Anal. Biochem.* **93:**257–260.

Hitchcock, P. J., and Brown, T. M., 1983, Morphological heterogeneity among Salmonella lipopolysaccharide chemotypes in silver stained polyacrylamide gels, *J. Bacteriol.* **154:**269–277.

Jackowski, G., and Liew, C. C., 1980, Fluorescamine staining of non-histone chromatin proteins as revealed by two-dimensional polyacrylamide gel electrophoresis, *Anal. Biochem.* **102**:321–325.

Marshall, T., and Williams, K. M., 1984, Artifacts associated with 2-mercaptoethanol upon high-resolution two-dimensional electrophoresis, *Anal. Biochem.* **139**:502–505.

Matheka, H. D., Enzmann, P. J., Bachrach, H. L., and Migl, B., 1977, *Anal. Biochem.* **81**:9–000.

Merril, C. R., and Goldman, D., 1984, Detection of polypeptides in two-dimensional gels using silver staining, in: *Two-Dimensional Gel Electrophoresis of Proteins: Methods and Applications* (J. E. Celis and R. Bravo, eds.), pp. 111–126, Academic Press, New York.

Merril, C. R., Goldman, D., and Van Keuren, M. L., 1982, Simplified silver protein detection and image enhancement methods in polyacrylamide gels, *Electrophoresis* **3**:17–23.

Merril, C. R., Switzer, R. C., and Van Keuren, M. L., 1979, Trace polypeptides in cellular extracts and human body fluids detected by two-dimensional electrophoresis and highly sensitive silver stain, *Proc. Natl. Acad. Sci. USA* **76**:4335–4339.

Nelles, L. P., and Bamburg, J. R., 1976, Rapid visualization of protein–dodecyl sulfate complexes in polyacrylamide gels, *Anal. Biochem.* **73**:522–531.

Ochs, D., 1983, Protein contaminations of sodium dodecyl sulfate polyacrylamide gels, *Anal. Biochem.* **135**:470–474.

Ochs, D. C., McConkey, E. H., and Sammons, D. W., 1981, Silver stains for protein in polyacrylamide gels: A comparison of six methods, *Electrophoresis* **2**:304–307.

Ragland, W. L., Pace, J. L., and Kemper, D. L., 1974, Fluorometric scanning of fluorescamine-labeled proteins in polyacrylamide gels, *Anal. Biochem.* **59**:24–33.

Sammons, D. W., Adams, L. D., and Nishizawa, E. E., 1981, Ultrasensitive silver-based color staining of polypeptides in polyacrylamide gels, *Electrophoresis* **3**:135–141.

Sammons, D. W., Adams, L. D., Vidmar, T. J., Hatfield, C. A., Jones, D. H., Chuba, P. J., and Crooks, S. W., 1984, Applicability of color silver stain (Gelcode® System) to protein mapping with two-dimensional gel electrophoresis, in: *Two-Dimensional Gel Electrophoresis of Proteins: Methods and Applications* (J. E. Celis and R. Bravo, eds.), pp. 111–126, Academic Press, New York.

Schetters, H., and McLeod, B., 1979, Simultaneous isolation of major viral proteins in one step, *Anal. Biochem.* **98**:329–334.

Switzer, R. C., Merril, C. R., and Shifrin, S., 1979, A highly sensitive silver stain for detecting proteins and peptides in polyacrylamide gels, *Anal. Biochem.* **98**:231–237.

Takagi, T., Kubo, K., and Isemura, T., 1977, Simple visualization of protein bands in SDS–polyacrylamide gel electrophoresis by the insoluble complex formation between SDS and a cationic surfactant, *Anal. Biochem.* **79**:104–109.

Tjissen, P., and Kurstak, E., 1979, A simple and sensitive method for the purification and peptide mapping of proteins solubilized from densonucleosis virus with sodium dodecyl sulfate, *Anal. Biochem.* **99**:97–104.

Wilson, C. M., 1979, Studies and critique of amido Black 10B, Coomassie Blue R, and Fast Green FCF as stains for proteins after polyacrylamide gel electrophoresis, *Anal. Biochem.* **9**:263–278.

Wilson, C. M., 1983, Staining of proteins on gels: Comparisons of dyes and procedures, *Methods Enzymol.* **91**:236–247.

Wray, W., Boulikas, T., Wray, V. P., and Hancock, R., 1981, Silver staining of proteins in polyacrylamide gels, *Anal. Biochem.* **118**:197–203.

Chapter 5

Basic Principles of Posttransitional Modification of Proteins and Their Analysis Using High-Resolution Two-Dimensional Polyacrylamide Gel Electrophoresis

I. INTRODUCTION

Many of the biological activities and physicochemical properties of proteins can be attributed to their posttranslational modifications, including glycosylation (the addition of sugar residues), sulfation (the addition of SO_4 groups), or phosphorylation (the addition of PO_4 residues). One of the most important uses of high-resolution two-dimensional polyacrylamide gel electrophoresis (PAGE) is for analysis of such modifications, which frequently cause shifts in the charge and thus the migration of proteins. It is therefore important to have a basic understanding of these modifications if the analysis methods described in this text are to be used.

II. PROTEIN GLYCOSYLATION

Many hormones such as follicle-stimulating hormone (FSH), immunoglobulins, and a number of enzymes are proteins that have now been shown to contain covalently bound sugars (Ashwell and Morell, 1974; Longmore and Schacter, 1982; Pazur and Aronson, 1972; Sharon, 1975). The use of high-resolution two-dimensional PAGE is extremely valuable in the analysis of these modifications. This technology can be combined with a number of other techniques in order to obtain a variety of important information on the chemical properties of specific proteins. In order to appreciate the effects that carbohydrates can have on the migration of proteins in electrophoretic fields, however, it is essential to understand the nature of these side chains that modify proteins.

A. Definitions and Basic Structures

The term *oligosaccharide* refers to carbohydrate polymers composed of two or more monosaccharide residues. Different monosaccharides may vary in the length of their carbon chain, although the most common sugars in glycoproteins consist of six carbons. Figure 5.1 illustrates the most common monosaccharides constituting glycoproteins. Each of these monosaccharides exists in one specific ring form that, in solution, adopts a preferred conformation such as the chair conformation (Rees, 1977; Hughes, 1983). It is important to note, however, that little is known about the conformations that are adopted by these monosaccharide units when they are incorporated into such complex structures as glycoproteins.

Glycoproteins are proteins that contain oligosaccharides covalently attached to specific amino acid residues. Because other complex glyco-conjugates such as proteoglycans and glycolipids are not generally re-solved with the gel systems outlined in this book, these are not discussed. The oligosaccharide side chains of glycoproteins are linked together by covalent bonds in which the hydroxyl of carbon atom 1 is bonded through the elimination of a water molecule, with an available hydroxyl on a carbon of the second monosaccharide other than C-1. The carbohydrate side chains of glycoproteins can range from disaccharides to complex struc-tures containing 18 monosaccharides or more and that can have extremely complex conformations. Some glycoproteins such as collagen and some proteins of submaxillary gland secretions contain disaccharide units, whereas other glycoproteins may contain five or more different monosaccharides. In fact, the number of carbohydrate units per protein molecule may vary from two to four to more than 300 (Spiro, 1973, 1976). Detailed analyses of many common carbohydrates in glycoproteins are given in Spiro (1973).

B. Protein–Carbohydrate Linkages

The unique structural feature of glycoproteins is the covalent sugar–amino acid bond, which attaches the carbohydrate units to the peptide chain. This linkage always involves the C-1 of the most internal sugar residue and a functional group of an amino acid within the peptide

A. Hexosamines

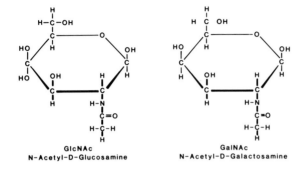

GlcNAc
N-Acetyl-D-Glucosamine

GalNAc
N-Acetyl-D-Galactosamine

B. Hexoses

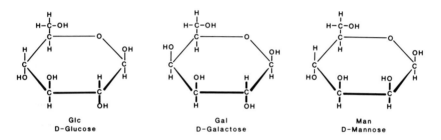

Glc
D-Glucose

Gal
D-Galactose

Man
D-Mannose

C. Deoxyhexose

D. Sialic Acid

Fuc
L-Fucose

NANA
N-Acetylneuraminic Acid

Figure 5.1. Common monosaccharides constituting oligosaccharide side chains of glycoproteins.

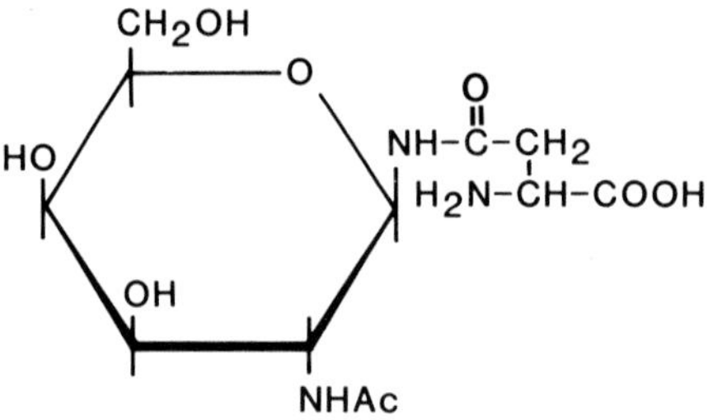

Figure 5.2. Structure of asparagine and N-acetylglucosamine linkage (2-acetamido-1-(L-β-aspartamido)-1,2-dideoxy-β-D-glucose) (N linkage).

chain. Rarely does a polypeptide chain of a glycoprotein contain a single unique carbohydrate sequence. Usually several and frequently distinct amino acid residues are glycosylated, with each site containing a mixture of similar but not necessarily identical oligosaccharide side chains. A limited number of amino acid residues with side chains can be involved in sugar attachment.

A common protein–carbohydrate linkage (N-linked) involves the amide group of the amino acid, asparagine, with the C-1 of N-acetylglucosamine, GlcNAc-Asn (Fig. 5.2). This N-acetylglucosamine, which joins the pep-

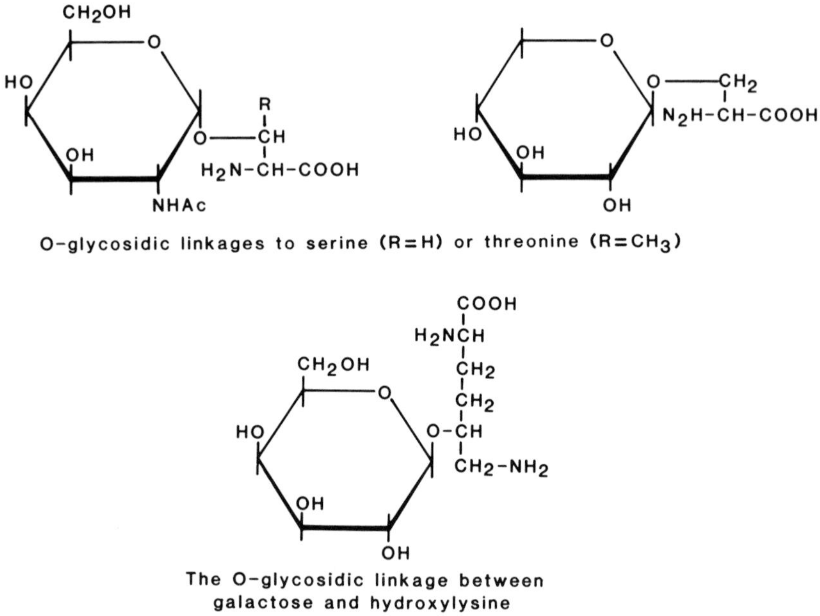

O-glycosidic linkages to serine (R=H) or threonine (R=CH₃)

The O-glycosidic linkage between galactose and hydroxylysine

Figure 5.3. Structure of linkages of sugars joined through C-1 by glycosidic bonds to the hydroxylated side chains of amino acids serine, threonine, and hydroxylysine (O linkage).

tide backbone through a glycosylamine bond, will therefore be the site of attachment of other monosaccharides.

The second class of (O-linked) linkage sugars (GalNAc-Ser or Thr) are joined through the last C on the carbohydrate to the hydroxylated side chains of threonine, serine, hydroxylysine or hydroxyproline by glycosidic bonds (see Fig. 5.3). In general, the glycoproteins of higher organisms most frequently contain the sugar N-acetylgalactosamine, which is linked to serine or threonine (O-glycans). The unique linkage found in vertebrate and invertebrate collagen consists of D-galactose joined to hydroxylysine via a β-glycosidic linkage and the most common linkage found in plant glycoproteins consists of L-arabinose glycosidically linked to hydroxyproline (see discussion by Hughes, 1983).

While a variety of other linkages have been described elsewhere, the goal of this chapter is simply to help the investigator appreciate the complexity of sugar side chains that give rise to glycoprotein charge heterogeneity, which is apparent in polyacrylamide gels. No attempt has been made to include detection methods for the sequence analysis of carbohydrate side chains. These methods have been outlined in detail elsewhere (Montreuil, 1980; Hughes, 1983) and generally require the extensive proteolysis of the proteins, leaving multiple small glycopeptides that are then analyzed by methods such as gel filtration and by gas–liquid chromatography (GLC) or high-pressure liquid chromatography (HPLC).

C. Synthesis and Processing of Glycoproteins

Studies have demonstrated that the peptide backbone of glycoproteins assembles on membrane-bound ribosomes and that most of the carbohydrate is incorporated following release of the peptide backbone from the ribosome (Schacter and Roden, 1973; see also review by Rothman, 1985). The biosynthesis of glycoproteins is a complex process that involves many steps.

1. Enzymes and Nucleotides in Glycoprotein Biosynthesis

Glycosyltransferases are enzymes that are directly involved in the synthesis of N- as well as O-glycans of glycoproteins. These glycosyl enzymes catalyze the reaction diagrammed in Figure 5.4; see also Table 5.1.

Sugar nucleotides, the α- and β-glycosyl esters of nucleotides, are the immediate donors of monosaccharides in the glycosylation procedure. The nucleotide moiety may be uridine, guanosine, or cytidine, which exist

$$X^* - \text{sugar} + \text{acceptor} \longrightarrow \text{Sugar-acceptor} + X$$

Figure 5.4. Reaction catalyzed by glycosyltransferases. *High-energy sugar nucleotide. The basic reaction is one in which a sugar unit is transferred from an activated high-energy derivative to a suitable acceptor that may be an asparagine, serine, threonine, or hydroxylysine residue in a polypeptide or a carbohydrate sequence in a glycoconjugate (Hughes, 1983). The individual glycosyltransferases are some of the most discriminating biologically active enzymes and exhibit very exact specificities toward the nature of the high-energy derivative as well as for the acceptor. The major sugar nucleotides involved in protein glycosylation are summarized in Table 5.1.

in sugar nucleotides as pyrophosphate esters except in sialic acid derivatives, in which a phosphodiester linkage between the ribose moiety of the nucleotide and the sialic acid is found. To date, more than 100 different sugar nucleotides have been identified. Most of these have the general structure XDP-sugar, where X can be any of the nucleosides (adenosine, guanosine, cytidine, uridine) and the sugar moiety can be a variety of structures. The monosaccharide moieties of the sugar nucleotides are transferred to form a particular glycosidic linkage. Because the enzymes involved in each reaction are highly specific, a wide range of glycosyltransferases are necessary to give rise to diverse carbohydrate chains.

Table 5.1. Summary of
Major Sugar Nucleotides Involved
in Protein Glycosylation[a]

Sugar	Activated form
Galactose	UDP-galactose
Mannose	GDP-mannose
N-Acetylglucosamine	UDP-GlcNAc
N-Acetylgalactosamine	UDP-GalNAc
Fucose	GDP-fucose
N-Acetylneuraminic acid	CMP-NeuNAc
Xylose	UDP-Xyl
Glucuronic acid	UDP-GlcUA

[a]From Hughes (1983).

2. Asparagine-Linked Oligosaccharides

The addition of sugar to protein may occur either directly from sugar nucleotides or indirectly through a lipid intermediate, which is based on the polyisoprenoid, dolichol (Waechter and Lennarz, 1976; Parodi and Leloir, 1976; Kornfeld *et al.*, 1978; Chapman *et al.*, 1979). During the formation of asparagine-linked oligosaccharides, the sugar chains are built on a lipid transferred to the polypeptide. Modification by removal or addition of monosaccharide residues follows. Figure 5.5 summarizes this processing. The first step of this process, which involves the synthesis of dolichol *N*-acetylglucosamine pyrophosphate, is inhibited by the drug tunicamycin. This drug is commonly used in studies using radiolabeled carbohydrate precursors to examine the synthesis of glycoproteins.

This glycosylation process has been shown to occur just after the nascent polypeptide is synthesized and is discharged into the cisternal space of the rough endoplasmic reticulum. This occurs after cleavage of the signal peptide and as the protein folds to expose segments of the polypeptide for glycosylation.

Two basic types of *N*-linked oligosaccharides have been described

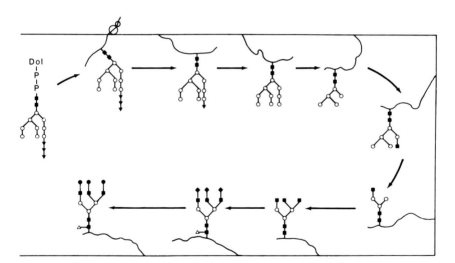

Figure 5.5. Proposed sequence for the processing of asparagine-linked oligosaccharides demonstrating the buildup of sugar chains on a lipid carrier molecule transfer to the polypeptide chain and modification by removal or addition of monosaccharides. (■) *N*-acetylglucosamine residues, (○) mannose residues, (▼) glucose residues, (●) galactose residues, (♦) sialic acid residues, and (△) fucose residues. (Modified from Kornfeld *et al.*, 1978, and reprinted with permission from the American Society of Biological Chemists, Inc.)

by Staneloni and Leloir (1979): (1) the high mannose type, and (2) the complex type (Fig. 5.6). In the high mannose type of oligosaccharides, the side chains have a variable number of mannose residues; in some cases, there may also be some acetylglucosamine residues. The side chains of the complex type contain galactose, *N*-acetylglucosamine, and sialic acid, although some may be incomplete in that they do not contain all carbohydrate residues.

3. O-Glycan Formation

The formation of *O*-glycans, as contrasted with *N*-glycans, has been shown to occur relatively late in biosynthesis (Hanover *et al.*, 1980). The addition of sialic acids, as well as the conversion of *N*-acetylneuraminic acid by hydroxylation and acetylation reactions, is associated with the smooth endoplasmic reticulum or Golgi membranes.

4. Nonenzymatic Glycosylation of Proteins

A variety of proteins, including albumin (Day *et al.*, 1979*a,b*), crystallin (Stevens *et al.*, 1978), hemoglobin (Winterhalter, 1981), insulin (Dolhofer and Wieland, 1979), and basic myelin protein (Fluckiger and Winterhalter, 1978) have been shown to be glycosylated by nonenzymatic mechanisms (see review by Fluckiger and Gallop, 1984). In the nonenzymatic glycosylation reaction, glucose and other hexoses react with unprotonated amino groups to form a Schiff base that can undergo a rearrangement to the corresponding 1-amino-1-deoxy-2-keto compound. In the case of hemoglobin, this process results in the decrease in positive charge caused by glycosylation of the *N*-terminal amino groups of the β-chain (Winterhalter, 1981).

Because this form of glycosylation is difficult to detect (Fluckiger and Gallop, 1984) and because artifacts of measurements have been reported (Trueb *et al.*, 1980), it is difficult at this time to assess the extent of occurrence of this type of glycosylation in complex protein systems.

D. Effects of Glycosylation on the Physicochemical and Biochemical Properties of Proteins

1. Effect on Solubility and Conformation of Proteins

Most proteins are active in an aqueous environment in which their hydrophobic residues tend to be forced to the interior of the structure,

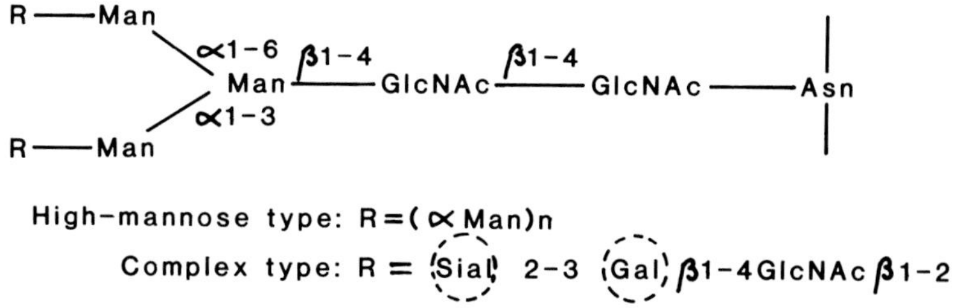

R —— Man

α1-6 β1-4 β1-4
Man —— GlcNAc —— GlcNAc —— Asn
α1-3

R —— Man

High-mannose type: R=(α Man)n

Complex type: R = Sial 2-3 Gal β1-4GlcNAc β1-2

Figure 5.6. Diagram of two basic types of N-linked oligosaccharides. (From Staneloni and Leloir, 1979.)

while their hydrophilic residues are exposed at the exterior of the molecule. The presence of carbohydrates, especially sialic acid, can significantly alter the charge of the protein and thus its solubility as well as its conformation. In general, glycosylation will increase the solubility of proteins (Berman and Lasky, 1985).

2. Effect on Glycoprotein Half-life

Glycosylation can render proteins more resistant to proteolysis (Schwarz et al., 1976; Older et al., 1978). This property would affect the half-life of the protein in vivo as well as interfere with proteolysis studies such as peptide mapping.

3. Effect of Glycosylation Immunogenicity and Antigenicity

Glycosylation is known to play an important role in the immunogenicity as well as the antigenicity of glycoproteins. The proper folding of a glycoprotein can alone alter its immunogenicity (Canton et al., 1982; Skehel et al., 1984). This alteration in antigenicity has been attributed to several factors: (1) the carbohydrate itself may be part of an antigenic determinant of the protein, (2) a nonglycosylated protein may express antigenic determinants that are not found on the glycosylated molecule, and (3) the attachment of carbohydrate may affect the overall folding of the protein and thus influence antigenic determinants at sites remote from the site of glycosylation (Berman and Lasky, 1985). The immunogenicity of a protein may be enhanced by carbohydrate simply because the immunogen may be protected from proteolysis, since glycosylation tends to render proteins more resistant to enzymatic digestion.

E. Carbohydrate Side-Chain Heterogeneity

When analyzing glycoproteins, the investigator should further be aware that the carbohydrate side chains may exhibit structural heterogeneity even within the same protein attached to identical amino acid residues or between different molecules of the same protein (see discussion by Spiro, 1973). For example, the glycoprotein, ovalbumin, has been reported to contain as many as six distinct glycopeptides (Huang *et al.*, 1970; Narasimhan *et al.*, 1980). This can account for some of the heterogeneity observed when these glycoproteins are analyzed with PAGE. The diversity exhibited by glycoproteins is often termed *microheterogeneity,* which occurs in glycoproteins containing a single carbohydrate unit as well as in proteins containing multiple units. In some glycoproteins, the heterogeneity of the carbohydrate units can be recognized in the intact molecule due to differences in charge (if sialic acid or sulfates are variants) or to differences in size if the carbohydrate makes up a substantial part of the total weight (Spiro, 1973).

It has been further established that microheterogeneity cannot be attributed to a genetic phenomenon (Cunningham and Ford, 1968; Kornfeld *et al.*, 1971). The proposed cause for heterogeneity of glycoprotein carbohydrate units is the differential processing of side chains during biosynthesis (Spiro, 1973). If completion of the carbohydrate units is not accomplished before the glycoprotein leaves the cell, molecules at various stages of processing would therefore be exported. It is further possible that attachment of a branch point containing a charged sugar such as sialic acid prior to completion of the main saccharide chain could impose steric hindrance to further action of glycosyltransferases (Spiro, 1973), therefore causing microheterogeneity. Whatever the cause, glycoprotein heterogeneity is a common property and one that can most easily be analyzed by two-dimensional PAGE.

F. Analysis of Glycoproteins Using Two-Dimensional Polyacrylamide Gel Electrophoresis

1. Charge and Molecular-Weight Heterogeneity

In general, glycoproteins can be better resolved by two-dimensional PAGE methods than they can with one-dimensional PAGE. Figure 5.7 illustrates glycoproteins that exhibit both extreme charge and molecular-weight heterogeneity in two-dimensional PAGE gels. Because the carbohydrate side chains may have charged residues such as sialic acid or sulfate, they may exhibit considerable charge heterogeneity and will not be resolved as a single charge species in isoelectric focusing (IEF). This heterogeneity will also affect the binding of the detergent SDS, resulting in apparent heterogeneity of molecular weight in the polyacrylamide gel. Since SDS is an anionic detergent, it binds to the hydrophobic regions of proteins and separates most of these into the component subunits (Weber and Osborn, 1969). This binding imparts a large negative charge to denatured randomly coiled polypeptides. Most multichain proteins bind approximately 1.4 g of SDS per gram protein and the resulting complexes behave as though they have a uniform negative charge and shape. Glycoproteins that contain acid-charged side chains, such as sulfates or sialic acid, may bind markedly different levels of SDS. This binding may therefore dramatically affect the mobility of the protein in the electrophoretic field as well as in the polyacrylamide matrix. For example, glycoproteins with high carbohydrate content, such as human erythrocyte glycoprotein and porcine ribonuclease, migrate at rates slower than would be expected from their molecular weight mass in SDS electrophoresis (Segrest et al., 1971). The low mobility of these glycoproteins was shown to be due to the reduced binding of detergent on a weight basis as compared with standard proteins of the same mass. In these studies, bovine serum albumin (BSA) was found to bind 0.73 g SDS per gram protein as compared with 0.38 g SDS per gram for the erythrocyte glycoprotein (60% carbohydrate) and 0.023 g SDS per gram of a proteolytic fragment of this glycoprotein (80% carbohydrate content).

It has further been demonstrated that the anomalous mobility of glycoproteins in SDS electrophoresis is most pronounced at low acrylamide concentrations (e.g., 5%), in which the sieving function of the gel matrix is the least (Segrest et al., 1971).

Since some glycoproteins are not readily solubilized by nonionic detergents, it is frequently necessary to use SDS as the solubilization reagent prior to IEF (see discussion in Chapter 3). Although most of the SDS will be removed from the proteins during IEF, allowing the protein to migrate to its appropriate position in the pH gradient, the degree and amounts of binding of SDS may affect the rate with which this happens for different glycoproteins. This could also bring about apparent charge heterogeneity in the IEF dimension.

In conclusion, it is apparent that while these methods are extremely useful in separating and resolving distinct glycoproteins, they are not always precise in allowing the precise determination of either the pI or molecular weights of many posttranslationally modified proteins.

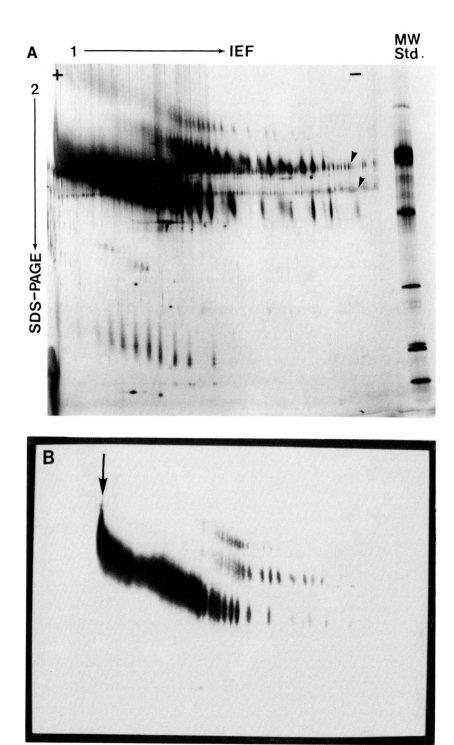

2. Identification of Glycoproteins by Lectin Binding and Polyacrylamide Gel Electrophoresis

Lectins are proteins of a nonimmune origin that will bind carbohydrates and cause precipitation of glycoproteins and can agglutinate cells. The activity of these proteins is inhibited by carbohydrates of a variety of specificities (Hughes, 1983). The first lectins identified were isolated from plants and have proved to be unique tools to study a variety of biochemical and biological phenomena (see review by Sharon and Lis, 1972). A list of commonly used lectins and their competitive sugars is given in Table 5.2.

Although these reagents have been useful tools, their use as chemical reagents in many studies have been overinterpreted due to a lack of understanding of the binding specificities of different lectins (Cederberg and Gray, 1979). Because lectins can have differential binding specificities, the binding to complex carbohydrates of glycoproteins may be very different from the binding specificity to purified monosaccharides (Irimura et al., 1975; Kornfeld et al., 1981; Midoux et al., 1980; Monsigny et al., 1980; Yamamoto et al., 1981).

G. Deglycosylation of Glycoproteins

To date, there is no single method that will effectively remove all carbohydrate side chains from complex glycoproteins. There are, however, a number of methods that will remove sufficient quantities of carbohydrates to reduce the microheterogeneity so that the proteins can be better analyzed by PAGE.

◄ ───

Figure 5.7. Protein pattern of sulfated, glycosylated proteins analyzed by two-dimensional polyacrylamide gel electrophoresis. (A) Protein pattern detected with the color-based silver stain Gelcode. The posttranslational modifications of these glycoproteins of the pig zona pellucida (the unique extracellular matrix surrounding the mammalian oocyte) results in extensive heterogeneity in polyacrylamide gels. (B) Immunoblot pattern obtained with polyclonal antibodies made against the protein species (arrow) obtained from a Coomassie Blue-stained two-dimensional gel. (From Skinner et al., 1984.) The major protein "family" has been shown to be composed of a protein that is differentially glycosylated and sulfated and exhibits a charge heterogeneity of approximately pH 4–9 and a molecular-weight heterogeneity of 40,000–120,000 (Dunbar et al., 1981). Molecular-weight markers (arrow, BioRad) were run in the second dimension along with the proteins separated in isoelectric focusing gel. Arrows represent silver stain artifacts.

**Table 5.2. Summary of Lectins Commonly Used in Biological
and Biochemical Studies**[a]

Agglutinin or lectin form	Sugar specificity of competitive sugars[b]
Concanavalin A (Con A)	α-D-man > α-D-Glc > α-GlcNAc
Vicia FABA (VFA)	α-D-Glc, α-D-man
Lens culinaris 1 CH (A + B) (LCH) (different from Con A in that fucose in the mannose-containing oligosaccharide chain is an important determinant)	α-D-man > α-D-Glc
Triticum vulgaris (WGA)	β-D-GlcNAc (1→4)-β-D GlcNAc₃ > (1→4)-β-D GLcNAc₂ > sialic acid
Phaseolus vulgaris (PHA-P)	β-Gal (1→4)-β-D
Ulex europaeus (UEA-II)	α-L-fuc(1→2)-D-Gal β-(1→4)D-GlcNaC (1→6)→2-fucosyl-lactose > (D-GlcNAc)₂
Dolichos biflorus (DBA)	α-D-GallNac
Glycine max (SBA)	α-D-GalNAc > β-D-GalNaC
Helix aspersa (HAA)	α-D-GalNAc ≫ α-D-GlcNAc
Riccinus communis (I and II) (RCA)	β-D-Gal > α-D-Gal
Trichosanthes kinlow II (TKA)	D-Gal
Viscum album (SRA, VAA)	D-Gal
Griffonia simplicifolia I (GSAI)	α-D-Gal > α-D-GalNAc
Arachis hypogaea (PNA)	D-Gal-β-(1→3)-GalNAc
Ulex europaeus (I and II)	α-L-Fucose (D-GlcNAc)₂
Limax flavus Aggl (LFA)	Sialic acid
Limulus polyhemus (LPA)	Sialic acid

[a]From Doyle (1982).
[b]Note that the specificity of the competitive sugar may be different from that of the complex sugar of the glycoconjugate.

1. Enzymatic

a. Exoglycosidases. Exoglycosidases are enzymes that cleave mono-saccharides from glycosidic linkages only when the sugars are terminal residues. Most of these enzymes will display a great specificity for the sugar moiety as well as the configuration of the glycosidic linkage (Kobata, 1979; Hughes, 1983). One enzyme that is commonly used to remove terminal sialic residues is neuraminidase. It is also common to use stepwise treatment with different enzymes. This is done because many of these enzymes are stearically prevented from reaching their appropriate substrate. For example, treatment of intact bacterial glycoproteins with ga-lactosidase has no effect on galactose removal. However, following treatment with neuraminidase to remove terminal sialic acid residues, more than 80% of galactose can be removed by galactosidase treatment (Hughes, 1983).

The loss of sialic acid from proteins, removed with neuraminidase, can have a marked effect on the isoelectric point of a protein. For example, the selective removal of 13–14-sialic residues from the carbohydrate side chains of a glycoprotein can result in a shift in the pI from 3.3 to 5.2 (Spiro, 1960).

b. Endoglycosidases. Endoglycosidases are enzymes that hydrolyze internal linkages and release oligosaccharides from glycoproteins or glycopeptides by proteolysis. While many such enzymes have been identified, the characterization of these enzymes has been dependent on the availability of oligosaccharides or glycopeptides of known structure (Hughes, 1983). Whereas many of these enzymes are helpful in the analysis of carbohydrate side chains, they are frequently inadequate for use in removing large quantities of carbohydrates from the protein backbone. These limitations are primarily due to the inaccessibility of the enzyme to intact molecules.

Recently, an enzyme (endoglycosidase F) was described that cleaves both complex and high-mannose glycans from asparagine-linked glycoproteins (Elder and Alexander, 1982; Plummer *et al.,* 1984). The enzyme has the added advantage of being active in the presence of nonionic detergents, and it works effectively after reduction and alkylation of proteins. (This enzyme has become increasingly popular and is now commercially available from Boehringer Mannheim Biochemicals and New England Nuclear.) See the method described in Appendix 10.

A major concern with the use of any enzyme preparation is that they are frequently not purified to homogeneity and may contain additional enzymes, including proteases. It is therefore necessary to interpret data obtained from these studies with caution. For example, the reduction of the apparent molecular weight of a protein may be due to partial proteolysis rather than to removal of significant quantities of carbohydrates.

2. Chemical Deglycosylation

a. β Elimination. The β-elimination reaction is one that is dependent on the lability of *O*-glycosidic linkages (e.g., GalNAc-Ser, GalNAc-Thr, or Xyl-Ser). These linkages will usually be cleaved by 0.05–0.1 M NaOH at room temperature within 24 hr. This reaction, however, is generally

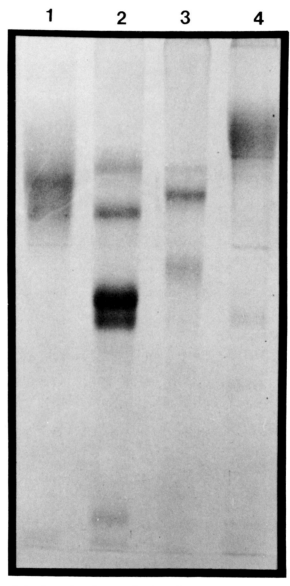

Figure 5.8. Analysis of complex glycoproteins by sodium dodecyl sulfate–polyacrylamide gel electrophoresis before and after deglycosylation using trifluoromethanesulfonic acid. *Well 1:* Pig zona pellucida glycoproteins. Extensive posttranslational modification makes resolution of distinct polypeptides impossible (compare this with two-dimensional pattern of these proteins shown in Fig. 5.6). *Well 2:* Pig zona pellucida glycoproteins treated with trifluoromethanesulfonic acid. While not all the carbohydrates were removed, sufficient quantities were removed to resolve the three major glycoproteins by one-dimensional PAGE. *Well 3:*

not quantitative, and methods have to be optimized for every glycoprotein. If alkali treatment is used with reducing conditions, the peptide-
linked monosaccharide will be converted to the corresponding sugar alcohol. The use of radioactive NaBH$_4$ (sodium borohydride) may then
facilitate the identification of the newly formed amino acids and sugar
alcohols (see discussion by Sharon, 1975).

It has further been demonstrated that both N- and O-linked oligosaccharides can be released using relatively mild conditions (Ogata and
Lloyd, 1982). Although these methods have the advantage that they can
be used for the analysis of carbohydrate moieties, the major limitation of
the methods is that the protein can be markedly altered. Nevertheless, it
is possible to use these methods in conjunction with other procedures to
gain important information about glycoproteins. One such study reports
labeling cell-surface proteins with [3H]glucosamine followed by isolation
of a specific antigen with a monoclonal antibody and analysis of the im-
munoprecipitated protein with SDS–PAGE. The radioactive protein was
detected by fluorography, cut from the gel, and then treated with alkaline
borohydride before analyzing carbohydrates by Sephadex chromatography (Ueda et al., 1981).

b. *Acid Hydrolysis of Glycosidic Bonds.* Hydrofluoric acid has been
used to remove carbohydrate from a variety of proteins, including peptide
hormones (Sairam and Schiller, 1979; Sairam and Fleshman, 1981; Chen
et al., 1982). Because this procedure requires special equipment and special handling of the hydrofluoric acid, it is not a method that can be easily
established in a laboratory and is not discussed in detail in this text.

Trifluoromethane sulfonic acid has been used to remove both O-
linked as well as N-linked carbohydrates from the protein backbone (see
Figs. 5.8 and 5.9). While the efficiency of this method varies depending
on the nature of the specific glycoproteins, it can be used to reduce
significantly the heterogeneity of glycosylated proteins in PAGE. If deglycosylated proteins are analyzed by high-resolution two-dimensional
PAGE in conjunction with lectin-blotting methods, the degree of degly-

---

Rabbit zona pellucida glycoproteins treated with trifluoromethane sulfonic acid. *Well 4:*
Rabbit zona pellucida glycoproteins. As with the pig zona pellucida glycoprotein, the sulfated
glycosylated forms of these proteins cannot be resolved by one-dimensional PAGE but the
deglycosylated gel proteins can be resolved. *Note:* This method may not remove all carbohydrate on all proteins. High-resolution two-dimensional PAGE and lectin blotting can
be used to further evaluate the presence of small amounts of carbohydrate on protein.
(Photograph from Skinner and Dunbar, 1986.)

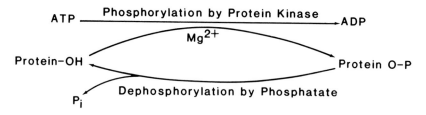

Figure 5.9. Diagram of protein phosphorylation reaction.

cosylation can be analyzed in greater detail. As this is a stray acid treatment, the protein may be altered and extreme care must be taken so that protein will not be hydrolyzed.

III. PROTEIN PHOSPHORYLATION

A. Introduction

Protein phosphorylation is another common form of posttranslational modification. Protein phosphorylation–dephosphorylation has been implicated in the regulation of numerous intracellular processes, including glycogenesis, gluconeogenesis, glycolysis, lipolysis, nucleic acid transcription, protein translation, and contractile protein interactions (for reviews, see Krebs and Beavo, 1979; Ochoa and deHaro, 1979; Stull, 1980; Manning *et al.*, 1980; Tash *et al.*, 1985). The enzymatic reaction resulting in the introduction of a phosphate group into a protein as shown in Figure 5.9 may result in marked changes in the biochemical as well as the biological properties of molecules. These changes could therefore regulate a variety of metabolic pathways. In order to better understand methods used in phosphorylation studies, the common terms used in protein phosphorylation studies are summarized in Table 5.2.

Cellular proteins contain three phosphorylhydroxy amino acids: phosphoserine, phosphothreonine, and phosphotyrosine. Other phosphorylated residues identified in proteins include thioesters of cysteine, phosphoamidates of lysine and histidine, and acid anhydrides of glutamic and aspartic acid. While some of these are intermediates in enzymatic reactions, others are stable modifications. Of all these potential phosphate esters of natural amino acids, however, only the three phosphorylhydroxy amino acids have been shown to exhibit a reasonable chemical stability to extremes of pH (see review by Cooper *et al.*, 1983).

Many intracellular events are regulated by second messengers such as cyclic adenosine monophosphate (cAMP) and calcium. Many of these involve, in part, the stimulation of protein phosphorylation, which commonly results in the activation of enzymes (see reviews by Tash *et al.*, 1985; Dedman *et al.*, 1979; Greengard, 1979; Means *et al.*, 1980). Recently, two-dimensional PAGE has become one of the most important methods for use in analyzing protein phosphorylation. In order to design experiments to analyze phosphoproteins, it is necessary to understand the mechanisms that regulate protein phosphorylation.

B. cAMP-Dependent Protein Phosphorylation

An enzyme first described by Krebs (1968), cAMP-dependent protein kinase, phosphorylates protein substances by catalyzing the transfer of the terminal phosphate of adenosine triphosphate (ATP) to serine residues of acceptor proteins. This enzyme is composed of two dissimilar subunits, regulatory (R) and catalytic (C), which physically dissociate for activity following binding of cAMP to the regulatory subunit:

$$R_2C_2 + 4\ cAMP \rightarrow R_2 \cdot cAMP_4 + 2C$$

Following dissociation, two active catalytic subunits express enzymatic activity and are capable of phosphorylating other proteins. The precise methods for assaying cAMP binding and protein kinase activities have been outlined in detail elsewhere (Tash *et al.*, 1985) and are not covered here. However, if one is interested in establishing that a protein is a physiological substrate of cAMP-dependent protein kinase in an experimental system, several criteria (outlined in Table 5.3) must be met to establish this.

A number of tissues have now been shown to contain multiple forms of cAMP-dependent protein kinases. Multiple forms can be resolved by analytical DEAE-column chromatography because the two holoenzymes comprise different regulatory subunits that differ in charge and molecular weight (see discussion by Tash *et al.*, 1985; Jahnsen *et al.*, 1985).

C. cAMP-Independent Protein Phosphorylation

In addition to the cAMP-dependent protein kinases, it has been demonstrated that a number of tissues contain multiple forms of both cAMP-independent as well as cAMP-dependent protein kinases (Corbin *et al.*,

**Table 5.3. Summary of Terms and Reagents Commonly Used in Protein
Phosphorylation Studies**

Reagent	Function
cAMP	Intracellular messenger involved in the regulation of cell function by mechanisms that stimulate protein phosphorylation
cAMP-dependent protein kinase	Phosphorylates protein substrates by catalyzing the transfer of the terminal phosphate of ATP to serine residues of acceptor probes
Adenosine triphosphatases (ATPases)	Enzymes that degrade or use ATP
Phosphodiesterases	cAMP-degradative enzymes
Phosphatases	Enzymes that remove phosphate groups from proteins
Sodium fluoride (NaF)	ATPase inhibitor (high concentrations may also inhibit phosphorylation reaction)
Methylisobutylxanthine	Phosphodiesterase inhibitor
8-bromo-cAMP	Analogue that will activate protein kinase but that is not degraded by phosphodiesterase
$[\gamma\text{-}^{32}P]$-ATP	Commonly used to label phosphoproteins in disrupted cell preparations
$[^{32}P]$orthophosphoric acid	Label that can and will be incorporated into the γ position of ATP of the endogenous pool of the cell (since ATP cannot transverse cell membrane)
8-azido-cAMP derivatives	Derivatives that bind to cAMP protein kinases and covalently link to regulatory subunit by azido side group

1975; Traugh and Traut, 1974). The cAMP-dependent and -independent
protein kinase activities can be separated by the use of phosphocellulose
column chromatography, and they can be further differentiated according
to their substrate specificities. For example, cAMP-dependent protein
kinases prefer histones (particularly histone H2B), whereas cAMP-independent protein kinases prefer either casein or phosvitin. [It is important
to note that conclusive differences in substrate specificities should be
validated by measuring both V_{max} and K_m of the enzymatic reaction (Tash
et al., 1985).]

D. Experimental Identification of cAMP-Dependent Protein Kinases Using Polyacrylamide Gel Electrophoresis

The recent development of derivatives of adenine nucleotides (e.g.,
8-azido derivatives) has made the analysis of cAMP-dependent protein

kinases easier, since these analogues can be used to determine qualitatively—but not quantitatively—whether a tissue contains one or both types of cAMP-dependent protein kinases (Tash *et al.*, 1985). These analogues will bind to a protein kinase in an identical manner as cAMP and will then be covalently linked to the regulatory subunit by the azido side group. The use of ^{32}P-labeled derivatives therefore allows for the analysis of these proteins by one- and two-dimensional PAGE (Tash *et al.*, 1985; Richards and Rolfes, 1980; Richards *et al.*, 1983, 1984; Jahnsen *et al.*, 1985) (see method outlined in Appendix 6).

E. Experimental Identification of Endogenous Substrates for Protein Kinases

While many phosphoproteins are readily identifiable by isotopic labeling, it is frequently more difficult to determine the precise functions of endogenous phosphorylation (see Table 5.4). Furthermore, experiments designed to analyze endogenous phosphorylation may require variations of conditions such as pH or ionic concentration. Because protein degradation or denaturation may also occur in complex protein samples, phosphorylation sites may be exposed that are not normally available *in vivo*. Likewise, endogenous phosphatase activities in cells may result in the removal of phosphate prior to analysis of samples.

Two approaches outlined by Tash *et al.*, (1985) are commonly used to identify and analyze endogenous phosphoproteins: (1) living cells can be labeled with [^{32}P]orthophosphoric acid; or (2) cell homogenates, extracts, or cytoskeletal preparations can be labeled with [γ-^{32}P]-ATP. The major advantage of ^{32}P-labeling of total cells is that protein denaturation problems are reduced. The disadvantage of this method, however, is that

Table 5.4. Criteria for Determination That a Protein Is a Physiological Substrate of cAMP-Dependent Protein Kinase[a,b]

1. The cell type under study contains a cAMP-dependent protein kinase.
2. A protein substrate exists that has a functional relationship to the cAMP-mediated response.
3. cAMP-dependent protein kinase is activated in response to hormone or stimulus.
4. Phosphorylation of the protein substrate alters its function *in vitro*.
5. Protein substrate is modified *in vivo* in response to cAMP-mediated stimulus.
6. A phosphoprotein phosphatase exists to reverse the phosphorylation process.

[a]From Tash *et al.* (1985).
[b]While many substrates will not meet all six criteria, at least five should be satisfied.

it is difficult to evaluate rapid changes in protein phosphorylation in response to treatment with hormones, drugs, and other agents. Also, it is impossible to determine the effects of ion on cAMP fluxes or the phosphorylation state of proteins. The labeling of cell homogenates or fractions, however, allows for more precisely controlled experimental conditions, permitting the evaluation of rapid effects of ions on cAMP or phosphorylation. This method is limited because artifactual phosphorylation due to alterations in proteins or to phosphatase activities must be taken into account (Tash *et al.*, 1985; see also Appendix 6).

F. Tyrosine-Specific Protein Kinases

A number of tyrosine-specific protein kinases have also now been described. Among these are the Abelsen murine leukemia virus (Witte *et al.*, 1980; Wang and Baltimore, 1983), epidermal growth factor (EGF) receptor (Cohen *et al.*, 1982; Ushiro and Cohen, 1980), and platelet-derived growth factor. As with other protein phosphorylation studies, polyacrylamide gels are also commonly used in these experiments.

REFERENCES

Ashwell, G., and Morell, A., 1974, The role of surface carbohydrates in the hepatic recognition and transport of circulating glycoproteins, *Adv. Enzymol.* **41**:99–128.

Berman, P., and Lasky, L. A., 1985, Engineering glycoproteins for use as pharmaceuticals, *Trends Biochem.* **3**(2):51–53.

Canton, A. J., Brownlee, G. G., Yewdell, J. W., and Gerhard, W., 1982, The antigenic structure of the influenza virus A/PR/8/34 hemagglutinin (H1 subtype), *Cell* **31**:417–427.

Cederberg, B. M., and Gray, G. R., 1979, *N*-acetyl-D-glucosamine binding lectins: A model system for the study of binding specificity, *Anal. Biochem.* **99**(1):221–230.

Chapman, A., Li, E., and Kornfeld, S., 1979, The biosynthesis of the major lipid-linked oligosaccharide of Chinese hamster ovary cells occurs by the ordered addition of mannose residues, *J. Biol. Chem.* **254**(20):10243–10249.

Chen, H. C., Shimohigashi, Y., Dufau, M. L., and Catt, K. J., 1982, Characterization and biological properties of chemically deglycosylated human chorionic gonadotropin, *J. Biol. Chem.* **257**:14446–14452.

Cohen, J., Fava, R. A., and Jawyer, S. T., 1982, Purification and characterization of epidermal growth factor receptor/protein kinase from normal mouse liver, *Proc. Natl. Acad. Sci. USA* **79**:6237–6241.

Cooper, J. A., Sefton, B. M., and Hunter, T., 1983, Detection and quantification of phosphotyrosine in proteins, *Methods Enzymol.* **99**:387–402.

Corbin, J. D., Keely, S. L., and Park, C. R., 1975, The distribution and dissociation of cyclic adenosine 3′, 5′-monophosphate-dependent protein kinases in adipose, cardiac, and other tissues, *J. Biol. Chem.* **250**:218–225.

Cunningham, L. W., and Ford, J. D., 1968, A comparison of glycopeptides derived from soluble and insoluble collagens, *J. Biol. Chem.* **243**:2390–2393.

Day, J. F., Thorpe, S. R., and Baynes, J. W., 1979*a*, Non-enzymatically glycosylated albumin *in vitro* preparation and isolation from normal human serum, *J. Biol. Chem.* **254**:595–597.

Day, J. F., Thornburg, R. W., Thorpe, S. R., and Baynes, J. W., 1979*b*, Non-enzymatic glycosylation of rat albumin studies *in vitro* and *in vivo*, *J. Biol. Chem.* **254**:9394–9400.

Dedman, J. R., Brinkley, B. R., and Means, A. R., 1979, Regulation of microfilaments and microtubules by calcium and cyclic AMP, *Adv. Cyclic Nucleotide Res.* **11**:131–174.

Dolhofer, R., and Wieland, O. H., 1979, Preparation and biological properties of glycosylated insulin, *FEBS Lett.* **100**:133–136.

Doyle, R. J., 1982, *Biotechs: Lectins in Diagnostic Microbiology*, pp. 1–8, E. Y. Laboratories, Inc., San Mateo, California.

Dunbar, B. S., Liu, C., and Sammons, D. W., 1981, Identification of the three major proteins of porcine and rabbit zonae pellucidae by high-resolution two-dimensional polyacrylamide gel electrophoresis: Comparison with serum, follicular fluid, and ovarian cell proteins, *Biol. Reprod.* **24**:1111–1124.

Elder, J. H., and Alexander, S., 1982, Endo-β-N-acetylglucosaminidase F: Endoglycosidase from *Flavobacterium meningosepticum* that cleaves both high mannose and complex glycoproteins, *Proc. Natl. Acad. Sci. USA* **79**:4540–4544.

Fluckiger, R., and Gallop, P. M., 1984, Measurement of non-enzymatic protein glycosylation, *Methods Enzymol.* **106**:77–87.

Fluckiger, R., and Winterhalter, K. H., 1978, *Biochemical and Clinical Aspects of Hemoglobin Abnormalities*, pp. 205–214, Academic Press, New York.

Greengard, P., 1979, Cyclic nucleotides, phosphorylated proteins, and the nervous system, *Fed. Proc.* **38**:2208–2217.

Hanover, J. A., Lennarz, W. J., and Young, J. D., 1980, Synthesis of N- and O-linked glycopeptides in oviduct membrane preparations, *J. Biol. Chem.* **255**:6713–6716.

Huang, C., Mayer, H. E., Jr., and Montgomery, R., 1970, Microheterogeneity and paucidispersity of glycoproteins. Part 1. The carbohydrate of chicken ovalbumin, *Carbohydr. Res.* **13**:127–137.

Hughes, R. C., 1983, *Glycoproteins*, Chapman and Hall, London.

Irimura, T., Kawaguchi, T., Terao, T., and Osawa, T., 1975, Carbohydrate binding specificities of so called galactose-specific phytohemagglutinins, *Carbohydr. Res.* **39**:317–325.

Jahnsen, T., Lohmann, S. M., Walter, U., Hedin, L., and Richards, J. S., 1985, Purification and characterization of hormone-regulated isoforms of the regulatory subunit of Type II cAMP-dependent protein kinase from rat ovaries, *J. Biol. Chem.* **260**(29):15980–15987.

Kobata, A., 1979, Use of endo- and exo-glycosidases for structural studies of glycoconjugates, *Anal. Biochem.* **100**:1–14.

Kornfeld, R., Keller, J., Baenziger, J., and Kornfeld, S., 1971, The structure of the glycopeptide of human gamma G myeloma, *J. Biol. Chem.* **246**:3259–3268.

Kornfeld, S., Li, E., and Tabas, I., 1978, The synthesis of complex-type oligosaccharides, *J. Biol. Chem.* **253**(1):7771–7778.

Kornfeld, K., Reitman, M. L., and Kornfeld, R., 1982, The carbohydrate-binding specificity of pea and lentil lectins: Fucose is an important determinant, *Biol. Chem.* **256**:6633–6640.

Krebs, E. G., and Beavo, J. A., 1979, Phosphorylation-dephosphorylation of enzymes, *Annu. Rev. Biochem.* **48**:923–959.

Longmore, G. D., and Schacter, H., 1982, Product identification and substrate-specificity studies on the GDP-L-fucose:2-acetamido-2-deoxy-β-D-glucoside (FUC→ASN-Linked

GlcNAc)-6-α-fucosyltransferase in a Golgi-rich fraction from porcine liver, *Carbohydr. Res.* **100**:365–392.

Manning, D. R., DiSalvo, J., and Stull, J. T., 1980, Protein phosphorylation: Quantitative analysis *in vivo* and in intact cell systems, *Mol. Cell Endocrinol.* **19**:1–19.

Means, A. R., Dedman, J. R., Tash, J. S., Tindall, D. J., VanSickle, M., and Welsh, M. J., 1980, Regulation of the testis sertoli cell by follicle stimulation hormone, *Annu. Rev. Physiol.* **42**:59–70.

Midoux, P., Grivet, J. P., and Monsigny, M., 1980, Lectin–sugar interactions: The binding of 1-O-methyl-di-N-trifluoroacetyl-beta-chitobioside to wheat germ agglutinin, *FEBS Lett.* **120**(1):29–32.

Monsigny, M., Roche, A. C., Sene, C., Maget-Dana, R., and Delmotte, F., 1980, Sugar–lectin interactions: How does wheat germ agglutinin bind sialoglycoconjugates?, *Eur. J. Biochem.* **104**(1):147–153.

Montreuil, J., 1980, Primary structure of glycoprotein glycans. Basis for the molecular biology of glycoproteins, *Adv. Carbohydr. Chem. Biochem.* **37**:157–223.

Narasimhan, S., Harpas, N., Longmore, G., Carver, J. P., Grey, A. A., and Schacter, H., 1980, Control of glycoprotein synthesis. The purification by preparative high voltage paper electrophoresis in borate of glycopeptides containing high mannose and complex oligosaccharide chains linked to asparagine, *J. Biol. Chem.* **255**:4876–4884.

Nicholson, G. L., and Irimura, T., 1984, Estimating glycoprotein carbohydrate chain structures by lectin reactivities in polyacrylamide gels, *Biol. Cell* **51**(2):157–164.

Ochoa, S., and deHaro, C., 1979, Regulation of protein synthesis in eukaryotes, *Annu. Rev. Biochem.* **48**:549–580.

Ogata, S., and Lloyd, K. O., 1982, Mild alkaline borohydride treatment of glycoproteins— A method for liberating both N- and O-linked carbohydrate chains, *Anal. Biochem.* **119**(2):351–359.

Ogata, S., Muramatsuj, T., and Kobata, A., 1975, Fractionation of glycoproteins by affinity column chromatography on concanavalin A-Sepharose, *J. Biochem.* **78**:678–696.

Older, K., Pratt, R. M., and Yamada, K. M., 1978, Role of carbohydrates in protein secretion and turnover: effects of tunicamycin on the major cell surface glycoprotein of chick embryo fibroblasts, *Cell* **13**:461–473.

Parodi, A. J., and Leloir, L. F., 1979, The role of lipid intermediates in the glycosylation of proteins in the eukaryotic cell, *Biochim. Biophys. Acta* **559**(1):1–37.

Pazur, J. H., and Aroneon, N. N., 1972, Glycoenzymes: Enzymes of glycoprotein structure, *Adv. Carbohydr. Chem. Biochem.* **27**:301–341.

Plummer, J. H., Jr., Elder, J. H., Alexander, S., Phelan, A. W., and Tarentino, A. L., 1984, Demonstration of peptide: N-glycosidase F activity in Endo-B-N-acetylglucos-aminidase F preparations, *J. Biol. Chem.* **259**:10700–10704.

Rees, D. A., 1977, Polysaccharide Shapes, Chapman and Hall, London.

Reynolds, J. A., and Tanford, C., 1970, Binding of dodecyl sulfate to proteins at high binding ratios. Possible implications for the state of proteins in biological membranes, *Proc. Natl. Acad. Sci. USA* **66**:1002–1007.

Richards, J. S., and Rolfes, A., 1980, Hormonal regulation of cyclic AMP binding to specific receptor proteins in rat ovarian follicles. *J. Biol. Chem.* **255**:5481–5489.

Richards, J. S., Seghal, N., and Tash, J. S., 1983, Changes in content and cAMP dependent phosphorylation of specific proteins in granulosa cells of preantral and preovulatory ovarian follicles and corpora lutea, *J. Biol. Chem.* **258**:5227–5232.

Richards, J. S., Haddox, M., Tash, J. S., Walter, U., and Lohman, S., 1984, Adenosine 3′, 5′-monophosphate-dependent protein kinase and granulosa cell responsiveness to gonadotropins, *Endocrinology* **114**(6):2190–2198.

Rothman, J. E., 1985, The compartmental organization of the Golgi apparatus, *Sci. Am.* **253**(3):74–89.

Sairam, M. R., and Fleshner, P., 1981, Inhibition of hormone-induced cyclic AMP production and steroidogenesis in interstitial cells by deglycosylated lutropin, *Mol. Cell Endocrinol.* **22**:41–54.

Sairam, M. R., and Schiller, P. W., 1979, Receptor binding, biological, and immunological properties of chemically deglycosylated pituitary lutropin, *Arch. Biochem. Biophys.* **197**:294–301.

Schwarz, R. T., Rohrschneider, J. M., and Schmidt, M. F. G., 1976, Suppression of glycoprotein formation of Semliki Forest, influenza, and avian sarcoma virus by tunicamycin, *J. Virol.* **19**:782–791.

Segrest, J. P., Jackson, R. L., Andrews, E. P., and Marchesi, V. T., 1971, Human erythrocyte membrane glycoprotein: A reevaluation of the molecular weight as determined by SDS–polyacrylamide gel electrophoresis, *Biochem. Biophys. Res. Commun.* **44**:390–395.

Schacter, H., and Roden, L., 1973, The biosynthesis of animal glycoproteins, in: *Metabolic Conjugation and Metabolic Hydrolysis,* Vol. 3 (W. H. Fishman, ed.), pp. 1–149, Academic Press, New York.

Sharon, N., and Lis, H., 1972, Lectins: Cell-agglutinating and sugar-specific proteins, *Science* **177**:949–959.

Sharon, N., 1975, *Complex Carbohydrates: Their Chemistry, Biosynthesis and Functions,* Addison-Wesley, Reading, Massachusetts.

Skehel, J. J., Stevens, D. J., Daniels, R. S., Douglas, A. R., Knossow, M., Wilson, I. A., and Wiley, D. C., 1984, A carbohydrate side chain on hemagglutinins of Hong Kong influenza viruses inhibits recognition by a monoclonal antibody, *Proc. Natl. Acad. Sci. USA* **81**:1779–1783.

Skinner, S. M., and Dunbar, B. S., 1986, *Immunological Approaches to Contraception and Promotion of Fertility* (G. P. Talwar, ed.), Plenum Press, New York.

Skinner, S. M., Mills, T., Kirchick, H. J., and Dunbar, B. S., 1984, Immunization with zona pellucida proteins results in abnormal ovarian follicular differentiation and inhibition of gonadotropin-induced steroid secretion, *Endocrinology* **115**(6):2418–2432.

Spiro, R. G., 1960, Studies on fetuin, a glycoprotein of fetal serum. I. Isolation, chemical composition, and physicochemical properties, *J. Biol. Chem.* **235**:2860–2869.

Spiro, R. G., 1970, Glycoproteins, *Annu. Rev. Biochem.* **39**:599–638.

Spiro, R. G., 1973, Glycoproteins, in: *Advances in Protein Chemistry* (C. B. Anfinsen, J. T. Edsall, and F. M. Richards, eds.), pp. 350–467, Academic Press, New York.

Spiro, R. G., 1976, Isolation of fetuin, in: *Methods in Carbohydrate Chemistry,* Vol. VII (R. L. Whistler and J. N. BeMiller, eds.), pp. 163–167, Academic Press, New York.

Spiro, R. G., 1976, Isolation of glycopeptides from glycoproteins by proteolytic digestion, in: *Methods in Carbohydrate Chemistry,* Vol. VII (R. L. Whistler and J. N. BeMiller, eds.), pp. 185–190, Academic Press, New York.

Staneloni, R. J., and Leloir, L. F., 1979, The biosynthetic pathway of the asparagine-linked oligosaccharides of glycoproteins, *Trends Biochem.* **4**:65–67.

Stevens, V. J., Rouzer, C. A., Monnier, V. M., and Cerami, A., 1978, Diabetic cataract formation: Potential role of glycosylation of lens crystallins (non-enzymatic glycosylation/sulfhydryl oxidation), *Proc. Natl. Acad. Sci. USA* **75**:2918–2922.

Stull, J. T., 1980, Phosphorylation of contractile proteins in relation to muscle function, *Adv. Cyclic Nucleotide Res.* **13**:39–93.

Tash, J. S., Guerriero, V., and Means, A. R., 1985, cAMP and calcium-dependent protein kinases and phosphoprotein identification, in: *Laboratory Methods Manual for Hor-*

mone Action and Molecular Endocrinology, 9th ed. (W. T. Schrader and B. W. O'-Malley, eds.), pp. 10.1–10.50, Houston Biological Association, Houston, Texas.

Traugh, J. A., and Traut, R. R., 1974, Characterization of protein kinases from rabbit reticulocytes, *J. Biol. Chem.* **249:**1207–1212.

Trueb, B., Holenstein, C. G., Fisher, R. W., and Winterhalter, K. H., 1980, Non-enzymatic glycosylation of proteins, *J. Biol. Chem.* **255**(14):6717–6720.

Ueda, R., Ogata, S., Morrissey, D. M., Finstad, C. L., Szkudlarek, J., Whitmore, W. F., Oettgen, H. F., Lloyd, K. O., and Old, L. J., 1981, Cell surface antigens of human renal cancer defined by mouse monoclonal antibodies: Identification of tissue-specific kidney glycoproteins, *Proc. Natl. Acad. Sci. USA* **78:**5122–5126.

Ushiro, H., and Cohen, S., 1980, Identification of phosphotyrosine as a product of epidermal growth factor-activated protein kinase in A-431 cell membranes, *J. Biol. Chem.* **255:**8363–8365.

Walsh, D. A., Perkins, J. P., and Krebs, E. G., 1968, An adenosine 3′, 3′-monophosphate-dependent protein kinase from rabbit skeletal muscle, *J. Biol. Chem.* **243:**3763.

Waechter, C. J., and Lennarz, W. J., 1976, The role of polyprenol-linked sugars in glyco-protein synthesis, *Annu. Rev. Biochem.* **45:**95–112.

Wang, J. Y., and Baltimore, D., 1983, Characterization of the Abedsen murine leukemia virus encoded tyrosine-specific protein kinase, *Methods Enzymol.* **99:**373–378.

Weber, K., and Osborn, M., 1969, The reliability of molecular weight determinations by dodecyl sulfate–polyacrylamide gel electrophoresis, *J. Biol. Chem.* **244:**4406–4412.

Winterhalter, K. H., 1981, Determination of glycosylated hemoglobins, *Methods Enzymol.* **76:**732–739.

Witte, O. N., Dasgupta, A., and Baltimore, D., 1980, Abelson murine leukaemia virus protein is phosphorylated *in vitro* to form phosphotyrosine, *Nature (Lond.)* **283:**826–831.

Yamamoto, K., Tsuji, T., Matsumdo, I., and Osawa, T., 1981, Structural requirements for the binding of oligosaccharides and glycopeptides in immobilized wheat germ agglutinin, *Biochemistry* **20:**5894–5899.

Chapter 6

Isotopic Labeling of Proteins for Electrophoretic Analysis

I. INTRODUCTION

One of the most commonly used methods for high-resolution two-dimensional polyacrylamide gel electrophoresis (PAGE) has been analysis of proteins during their synthesis *in vivo* or *in vitro* in cell or tissue culture or in *in vitro* translation systems. These methods can also be used to analyze proteins undergoing posttranslational modification, e.g., glycosylation. The high specific activity of [^{35}S]methionine has made this a popular reagent for radiolabeling proteins and is a sensitive method used to analyze many proteins in a variety of experimental systems. Other radiolabeled amino acids or mixes of several of these have also been used successfully to analyze synthesis of proteins. As with other methods of protein identification, isotopic labeling methods have certain advantages and limitations. It is important to appreciate these factors in all experimental designs.

II. AMINO ACID LABELING DURING PROTEIN SYNTHESIS

A. Advantages of Radiolabeling Proteins

While labeling of proteins during synthesis with radiolabeled precursors is technically a straightforward procedure, several factors must be taken into consideration. There are several advantages to these labeling methods when used in conjunction with the analysis of high-resolution two-dimensional PAGE:

1. The detailed analysis of proteins being actively synthesized can be carried out.

103

2. The relative rates of synthesis of multiple proteins can be analyzed simultaneously.
3. It is possible to analyze alterations in synthesis of proteins experimentally (e.g., cell differentiation, responses to drugs on hormone treatment).

B. Disadvantages of Radiolabeling Proteins

The major disadvantages to these methods include the following:

1. The analysis of constitutive proteins, many of which have slow rates of synthesis, cannot be easily analyzed. Furthermore, most proteins will have varying rates of protein synthesis, making direct comparisons of synthetic rates difficult without carrying out large numbers of experiments.
2. Many labeling procedures have to be carried out utilizing *in vitro* culture systems that may not mimic physiological responses that would be typical of cells *in vivo*.
3. The incorporation of the commonly used radioisotope, [^{35}S]methionine, is dependent on the number of methionine residues in a given protein. The quantitation of this method therefore relies on the measurement of methionine residues in amino acids, which may vary markedly from protein to protein. (This problem can be overcome by using a mixture of ^{14}C-radiolabeled amino acids, which are also available commercially.)
4. These methods require more technical manipulations, since gels have to be processed for autoradiography.
5. The cost of commercially produced radiolabeled isotopes can be considerable.

1. Choice of Radiolabeled Precursor

Radiolabeling of proteins with [^{35}S]methionine has been a method used regularly in the analysis of protein patterns by two-dimensional PAGE (O'Farrell, 1975; Garrels, 1979; Anderson, 1981; Duncan and McConkey, 1982; Bravo *et al.,* 1981*a;* Bravo and Celis, 1982; Ivarie and O'Farrell, 1978). The high specific activity of this labeled amino acid makes this a sensitive protein detection method and permits detection of labeled proteins following reasonably short exposures to autoradiographic films. It is important to take into account, however, that most commercially available preparations of [^{35}S]methionine contain dithiothreitol, or 2-mercaptoethanol in order to prevent oxidation of methionine to sulfoxide which will not function as an amino acid precursor of protein synthesis. Since such reducing agents can greatly alter viability and function of some cell populations, care should be exercised to remove or dilute out these reagents before labeling. The major disadvantage of this method is that the detection of proteins or quantification of proteins is dependent on the incorporation of a single radiolabeled amino acid precursor.

Radiolabeling of proteins with mixtures of ^{14}C-labeled amino acids (Bravo *et al.*, 1981*b*; Bravo and Celis, 1982) has the advantage that multiple amino acids will be incorporated; therefore, any protein deficient in one or more amino acids will still be detected. Theoretically this method is more optimal for protein quantitation, but in reality, it is generally more time consuming due to the lower specific activity of ^{3}H- or ^{14}C-amino acids and is more expensive than [^{35}S]methionine labeling.

2. Choice of Radiolabeling Methods

a. Labeling of Proteins in Vivo. Methods of labeling proteins using radiolabeled amino acid precursors have been used to label animal tissues *in vivo*. However, because the radiolabeled amino acid precursor is usually diluted rapidly in animal tissues and is competed for by endogenous amino acids, it is frequently difficult to obtain sufficient incorporation into proteins for this to be a generally applicable method. For this reason, most studies using radiolabeled amino acid precursors are carried out *in vitro*.

b. Labeling of Proteins in Vitro. Biosynthesis labeling of cells in tissue culture is a relatively straightforward process. As with other techniques, however, it is essential that the investigator appreciate the technical problems that can accompany this methodology. An excellent review and detailed method for the labeling of proteins for use in radiochemical sequence analysis has been outlined by Coligan *et al.* (1983).

In choosing the method of labeling, the major factors important in establishing experimental design are (1) the number of cells, and (2) the amount of radiolabel that can be used in an experiment. The cost of radiolabeled reagents as well as the level of detectability of a given isotope may also frequently affect the choice of labeling method.

The following factors should be taken into account to help investigators determine what information can be obtained for biosynthetically labeling proteins (Coligan *et al.*, 1983):

1. The incorporation of radiolabel will increase linearly with the concentration of the radioactive amino acid in the labeling medium

up to the point at which the specific activity of the amino acid in the labeling culture approaches the specific activity of the radiolabeled amino acid being added. For amino acids that have a high specific activity (e.g., ^3H- and ^{35}S-labeled amino acids), the intracellular pools of nonradioactive amino acids are large enough that this limitation is not a practical problem.

2. Incorporation of the radiolabel will increase proportionately with the number of cells used in the labeling study. It is important to note that the quantity of amino acid must be regulated according to the number of cells as well as the cell density used in the experiment. In general, if the number of cells is lowered, no loss of total counts incorporated into protein will be achieved as long as the isotope concentration is raised proportionately. Therefore, if the number of cells available is limited or if a high specific activity protein is needed, the same amount of radiolabeled material can be used with fewer cells (and smaller volume) to produce smaller amounts of protein containing the same level of radioactivity (Caligan et al., 1983).

3. The choice of cell culture system to be used will depend on the experimental question being addressed. In general, a steady-state condition is desired. This usually requires that the experiment be initiated when the culture is in the log phase of growth, so that the cell culture will remain in a steady state for a period of time greater than that required to complete the experiment. If this is not the case, the results will be compromised, because it will not be possible to determine whether results are due to experimental perturbation of the system or to the drastic physiological changes occurring when the culture makes its transition from the lag and stationary phase (Cooper, 1977).

If organ or tissue culture is used, it will be necessary to carry out labeling at a time when the tissue is healthy and when cell death is minimal. This may require analysis of protein synthesis throughout the labeling period. Although long-term tissue culture can be used successfully, it will be necessary to optimize culture conditions for your tissue.

C. Equilibrium, Pulse, and Pulse-Chase Labeling Studies

While methodology for the incorporation of radiolabeled amino acid precursors is technically straightforward, the design of experiments that will lead to meaningful results is frequently not adequate. A number of careful considerations must be taken into account in order to interpret results adequately as well as to obtain optimal information from experiments (Schimke, 1975). The general considerations for labeling proteins are similar as those previously outlined for studies to analyze nucleic acids (see discussions by Cooper, 1977).

Three basic types of labeling experiments can be carried out to analyze protein synthesis. These methods are commonly referred to as equilibrium, pulse, and pulse-chase labeling and are summarized in Table 6.1. Studies utilizing these methods can be designed to take into account protein degradation as well as protein synthesis.

1. Equilibrium Labeling

These experiments are carried out by incubating a cell or tissue in the presence of a large amount of low to moderate specific activity radioactive material and because long-term labeling is used for equilibrium labeling. The time of preincubation may vary from minutes to hours, depending on the experimental system. This is used to permit the appropriate amino acid precursor pools to reach constant specific activity. After this is achieved, the specific experiment can be initiated by adding the experimental variable of interest such as hormone or drugs. Two basic criteria should be observed when carrying out such an equilibrium study: (1) the experimental parameter to be studied should be analyzed only after the precursor amino acid pools are at a constant specific activity, and (2) the proteins being synthesized should possess a uniform distribution of label of constant specific activity throughout (see Section II.A.1.).

2. Pulse Labeling

Pulse labeling is used to monitor the rate of synthesis of a given molecule and is carried out by adding the radiolabeled amino acid pre-

Table 6.1. Summary of Protein-Radiolabeling Methods

Equilibrium labeling: Label is added to culture when it is at steady state and left throughout experiment.

Pulse labeling: Label is added for defined periods of time (usually short periods) during the synthesis study. The extent of labeling of a protein will depend on the duration of label as well as its rate of synthesis.

Pulse-chase labeling: Label is added for a defined period of time followed by removal of label and substitution with nonlabeled precursor (500–1000-fold greater concentration then radiolabeled precursor). (The extent of labeling will depend on duration of label as well as on its rate of synthesis.)

cursor at varying times throughout an experiment. As the duration of labeling period decreases to zero, the observed rate of synthesis will approach the true instantaneous rate of synthesis (Cooper, 1977). Since the labeling periods used here are generally for short periods of time, a precursor with a high specific activity is usually necessary. Furthermore, the period of incubation in the presence of the radioactive substrate will depend on the half-life of the molecule being labeled, and the duration of the labeling period should optimally be no more than 10% of the half-life values observed for the molecule being labeled (Cooper, 1977). Cooper (1977) also points out two critical factors that must be taken into account when carrying out pulse-labeling procedures: (1) the synthesis is occurring both before and during the labeling procedure, and (2) only the molecules or parts of the molecules synthesized during the labeling period are radioactive.

3. Pulse-Chase Labeling

The pulse-chase protein-labeling method is a variation of the pulse-labeling method in that the radioactive label is removed and replaced with 500–1000-fold of the nonreactive amino acid. The culture is then sampled at various times after the radiolabeled precursor is removed. This method can be used to get some information on precursor–product relationships (see discussion by Cooper, 1977).

D. Use of Protein Synthesis Inhibitors in Radiolabel Studies

In order to determine whether there are changes in amounts of protein in an experimental system, inhibitors of protein synthesis can be used as first approximation to answer this question. Table 6.2 summarizes com-

Table 6.2. Summary of Inhibitors Commonly Used in Protein Synthesis and Post-translational Modification Studies in Eukaryotic Cells

Inhibitor	Action
Puromycin	Inhibits protein synthesis by causing premature release of nascent polypeptide chains by its addition to growing chain end
Cyclohexamide	Inhibits protein synthesis by blocking the peptidyl transferase reaction on ribosomes
Tunicamycin	Inhibitor of N-linked glycosylation (can inhibit protein synthesis at high concentration)

mon inhibitors used in labeling studies. The limitation of results obtained from studies when inhibitors are used is that it is not possible to determine whether the interference is due to alteration in protein synthesis or to protein degradation, since both cases require new protein synthesis. It is further not possible to determine whether the synthesis of a specific protein is inhibited or whether other proteins are inhibited, in which case the synthesis or activity of other specific proteins is altered. Finally, the use of protein synthesis inhibitors can completely inhibit protein synthesis and cause cell or tissue death if used in high dosages for sufficient periods of time. These drugs are therefore useful primarily when changes in protein synthesis are relatively rapid.

E. Interpretation of Radiolabeling Experiments

1. Culture Conditions

It is necessary to take into account the status of the tissue or cells being labeled in these studies. For example, if cells in culture are being labeled, the stage of the cell cycle and condition of growth (e.g., logarithmic or confluency) may dramatically alter synthesis results. For many studies, it is preferable to use steady-state conditions.

If tissues or organ cultures are to be labeled, it will be necessary to optimize conditions to achieve uniform labeling throughout the tissue as well as an adequate gas mixture during incubation. (Examples of conditions that are generally adequate for labeling protein in cells or tissues are given in Appendix 6.)

2. Protein Degradation

Depending on the method of labeling to be used and the information desired from an experiment, protein degradation may or may not be critical. If short-term radiolabeling methods are used, such as pulse or pulse-chase methods, protein degradation may not be a major problem. If long-term equilibrium studies are to be used, however, protein degradation may be significant, leading to misinterpretation of results. Because different proteins will have different rates of degradation as well as different rates of synthesis, the information obtained from the simultaneous comparison of radiolabeled proteins analyzed by one- or two-dimensional PAGE may be limited. If information on the actual rates of synthesis and degradation of proteins are necessary, rigorous experiments should be carried out to analyze a specific protein. The methods have been outlined in detail by Schimke (1975) and using immunochemical methods.

3. Simultaneous Comparison of Multiple Radiolabeled Proteins by High-Resolution Two-Dimensional Polyacrylamide Gel Electrophoresis

The major limitations of analyzing proteins that have been radiolabeled with amino acid precursors is that (1) different proteins incorporate different numbers of amino acids and at different rates, and (2) different proteins have different rates of synthesis as well as degradation. It is therefore frequently possible only to evaluate relative levels of different proteins.

Because protein patterns obtained by metabolic labeling can be used to detect proteins that are actively being synthesized, the protein patterns obtained from these samples may be markedly different from those obtained when direct protein staining methods are used. For example, since many abundant cellular proteins may be slowly synthesized but accumulate in the cell due to slow rates of degradation, they may be predominant patterns of cell samples when analyzed by direct staining methods but may not be detected by radiolabel methods. On the contrary, some proteins may be rapidly synthesized and therefore incorporate radiolabel so that they are easily detected by autoradiographic analysis of protein samples but, because their rates of degradation are also rapid, sufficient levels may not build up in the cell so that direct protein-staining methods do not successfully detect the protein.

In summary, because different protein-staining methods stain different proteins differently and because different proteins incorporate different levels of radioactive precursors at different rates, it is frequently difficult to compare rates of protein synthesis directly by the simple analysis of polyacrylamide gels. To determine the rates of synthesis precisely, it is necessary to carry out detailed studies such as those described by Schimke (1975) or Chafouleas *et al.* (1981) in which immunoprecipitations with a specific antibody are used.

4. Analysis of Secreted versus Cellular Proteins

It is frequently necessary to determine which proteins are secreted by cells as opposed to those that are cellular proteins. Most such studies have been carried out using cell cultures in which a variety of experimental parameters can be studied. Most studies use radiolabel precursor molecules (e.g., amino acids or carbohydrates). Advances in sample preparation and the development of sensitive silver stains have also made it possible to analyze proteins in such studies.

Because secreted proteins may be less abundant, it is usually necessary to concentrate these so they can be analyzed to electrophoretic methods. Some concentration methods such as lyophilization result in the simultaneous increase in salt concentration, a factor that will interfere substantially in electrophoresis, especially isoelectric focusing. Other methods, such as dialysis concentration or centrifuge concentration, may result in substantial loss of proteins due to adherence to dialysis membranes or proteolysis during concentration. We have therefore developed an alternative method that will rapidly and easily concentrate cell culture media for analysis by PAGE (G. A. Maresh and B. S. Dunbar, manuscript in preparation; see method outlined in Appendix 2). This method uses Celite, which is diatomaceous earth that has a capacity to adsorb proteins nonspecifically. After adsorption of proteins to the Celite, the sample is centrifuged and a small amount of electrophoresis solubilization buffer is added to remove the proteins from the insoluble Celite. The protein concentration can then be measured by the method described by Ramagli and Rodriguez (1985) and the sample directly analyzed by a one- or two-dimensional PAGE. The secretory proteins (Fig. 6.1) obtained in this

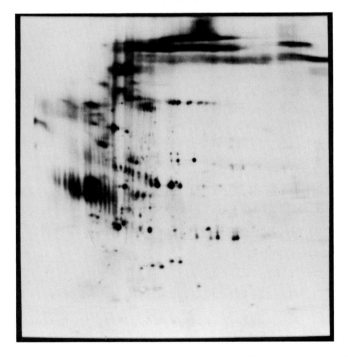

Figure 6.1. Autoradiograph of secretory protein pattern obtained from cells labeled with [^{35}S]methionine for 24 hr. The proteins were concentrated from the media using Celite. (From G. A. Maresh and B. S. Dunbar, unpublished observations.)

manner are typically distinct from the cellular protein patterns (Fig. 6.2) obtained from the cells.

The major problem associated with these studies is that proteins in the media may come from dying cells or from cells that are not attached to the culture dishes. The amount of cellular contamination can easily be determined by estimating the amount of the abundant cellular protein actin, which is easily identified by two-dimensional gel electrophoresis.

III. RADIOLABELING USING PROTEIN PHOSPHORYLATION METHODS

The chemical basis for protein phosphorylation has been discussed in detail in Chapter 5. The major reasons for using these types of radiolabeling methods are to (1) determine whether a protein is a substrate for cyclic adenosine monophosphate (cAMP)-dependent or -independent protein kinase or for a tyrosine-specific protein kinase, (2) determine which

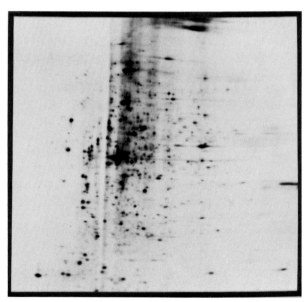

Figure 6.2. Autoradiograph of protein pattern of cells labeled with [^{35}S]methionine for 24 hr. Note the distinct protein spots (including actin) as compared with the secretory protein patterns obtained from the culture media of these cells (Fig. 6.1) in which apparent glycoproteins exhibit greater molecular and charge heterogeneity. (From G. A. Maresh and B. S. Dunbar, unpublished observations).

type of protein kinase or substrate will bind, (3) detect endogenous sub-strates for protein kinases, and (4) determine endogenous phosphoprotein in living cells or cell homogenates. The methods commonly used in these studies are discussed in Chapter 5 and outlined in detail in Appendix 6. High-resolution two-dimensional PAGE is an important method that can be used to identify substrates in these studies.

IV. CARBOHYDRATE LABELING OF GLYCOPROTEINS

In general, many of the same principles that apply to the radiolabeling of proteins also apply to the incorporation of radiolabeled sugar precursors into carbohydrate side chains of glycoproteins. Because radiolabeled sugar precursors are ^{3}H or ^{14}C labeled and because incorporation into specific carbohydrates may be limited, it may sometimes be difficult to incorporate sufficient numbers of radioactive labeled counts into glycoproteins for easy detection in proteins labeled *in vitro* as well as *in vivo*.

The choice of radiolabeled precursor to be used will depend on the nature of the carbohydrate side chain of the glycoconjugate (see discussion in Chapter 5). It may frequently be necessary to determine the carbo-hydrate composition in advance or to carry out pilot studies to determine which precursor will give optimal results (Otto and Muhlradt, 1980). Pre-cursors commonly used to label cells *in vitro* include (1) [^{3}H]glucosamine (Taylor and Weintraub, 1985; Liu and Jackson, 1985), (2) [^{3}H]mannose (Bradshaw *et al.*, 1985; Bradshaw and White, 1985*a,b*), and (3) L-[^{3}H]fucose (Honza *et al.*, 1983).

Because tunicamycin can inhibit N-linked protein glycosylation, it is commonly used in labeling studies to determine whether inhibition of glycosylation will alter the biological properties of synthesized proteins (Geetha-Habib and Bouck, 1982; Nolan and Farrell, 1985; Bradshaw and White, 1985*b*; Ercolani *et al.*, 1984; Savage and Baur, 1983; Bartlett *et al.*, 1984). Care must be taken in these studies to determine that levels of tunicamycin are optimized to ensure that inhibition of glycosylation occurs without interference with protein synthesis.

V. LABELING OF PROTEINS BY CHEMICAL MODIFICATION

To date, a large number of methods have been developed for the isotopic labeling of proteins. These methods have been summarized in detail by Hames and Rickwood (1981). When choosing a labeling method, it is important to understand the mechanisms of the labeling procedure. Properties of different labeling procedures that must be considered include (1) changes in physical and chemical properties of protein (e.g., charge properties), (2) alterations in biological activity, and (3) accessibility of proteins and/or amino acid residue in cells or in complex protein mixtures.

In general, labeling reagents can be classified according to the reactive groups of amino acids. These reactive groups include (1) the ($-NH_2$) of N-terminal or lysine residues of free amino groups, (2) thiol groups ($-SH$) of cystein residues, (3) the phenolic hydroxyl groups of tyrosine residues, (4) the imidazol groups of histidine residues, and (5) the aliphatic hydroxyl groups ($-CH_2OH$) of serine or threonine residues. Because most of these radiolabeling methods will alter the charge on proteins, alterations in the electrophoretic mobility in isoelectric focusing or binding of detergent used in molecular-weight separation methods may occur. For these reasons, most of these methods are not outlined in this text.

A method that is commonly used to label proteins isotopically, however, is one that does not dramatically alter the charge properties of proteins and is that of reductive methylation of free amino groups (Biocca et al., 1978). This method is outlined in detail in Appendix 6. This method uses formaldehyde to react with the amino group to form a Schiff base, which is then reduced with borohydride. If tritiated sodium borohydride is used, a N-[^{3}H]methyl derivative is produced (see Fig. 6.3).

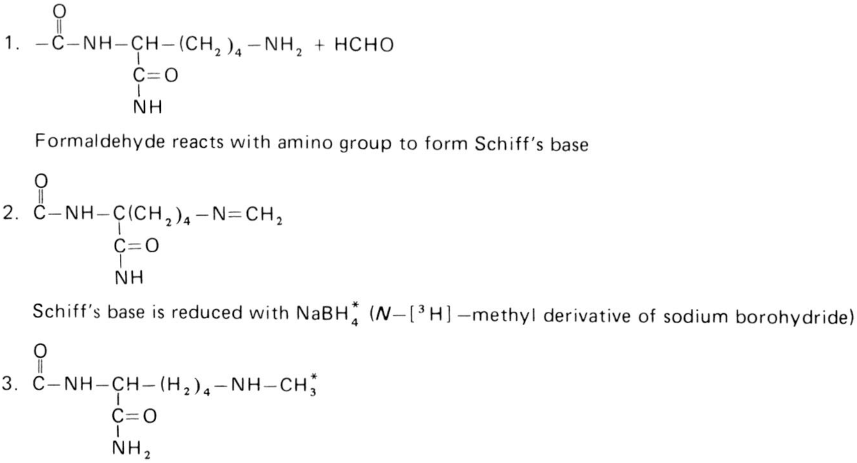

1. $$-\overset{\overset{\text{O}}{\|}}{\text{C}}-\text{NH}-\underset{\underset{\text{NH}}{\overset{|}{\underset{\overset{|}{\text{C}=\text{O}}}{}}}}{\text{CH}}-(\text{CH}_2)_4-\text{NH}_2 + \text{HCHO}$$

Formaldehyde reacts with amino group to form Schiff's base

2. $$\overset{\overset{\text{O}}{\|}}{\text{C}}-\text{NH}-\underset{\underset{\text{NH}}{\overset{|}{\underset{\overset{|}{\text{C}=\text{O}}}{}}}}{\text{C}}(\text{CH}_2)_4-\text{N}=\text{CH}_2$$

Schiff's base is reduced with NaBH$_4^*$ (N–[^{3}H] —methyl derivative of sodium borohydride)

3. $$\overset{\overset{\text{O}}{\|}}{\text{C}}-\text{NH}-\underset{\underset{\text{NH}_2}{\overset{|}{\underset{\overset{|}{\text{C}=\text{O}}}{}}}}{\text{CH}}-(\text{H}_2)_4-\text{NH}-\text{CH}_3^*$$

Figure 6.3. Diagram of procedure using sodium borohydride (^{3}H) and formaldehyde in the reductive methylation of free amino groups reaction for labeling proteins. *Radiolabeled group. (See procedure outlined by Biocco et al., 1978.)

REFERENCES

Anderson, N. L., 1981, Studies on gene expression in human lymphocytes using high-resolution two-dimensional electrophoresis, in: *Electrophoresis '81*, pp. 309–316, Walter deGruyter, Berlin.

Bartlett, F. J., French, F. S., and Wilson, E. M., 1984, *In vitro* synthesis and glycosylation of androgen-dependent secretory proteins of rat dorsal prostate and coagulating gland, *Prostate* 5(1):75–91.

Biocca, S., Calissano, P., Barra, D., and Fasella, P. M., 1978, Preparation of highly ^{3}H-labeled S-100 protein under non-denaturing conditions, *Anal. Biochem.* 87:334–342.

Bradshaw, J. P., and White, D. A., 1985a, The effect of tunicamycin or transferrin synthesis and secretion in hormonally stimulated explants of rabbit mammary gland, *Biosci. Rep.* 5(3):229–236.

Bradshaw, J. P., and White, D. A., 1985b, Identification of a major glycosylated protein of rabbit mammary gland and its appearance during development *in vivo*, *Int. J. Biochem.* 17(2):175–185.

Bradshaw, J. P., Hatton, J., and White, P. A., 1985, The hormonal control of protein-N-glycosylation in the developing rabbit mammary gland and its effect upon transferrin synthesis and secretion, *Biochim. Biophys. Acta* 847(3):344–351.

Bravo, R., and Celis, J. E., 1982, Updated catalogue of HeLa cell proteins: Percentages and characteristics of the major cell polypeptides labeled with a mixture of 16 ^{14}C-labeled amino acids, *Clin. Chem.* 28(4):766–781.

Bravo, R., Bellantin, J., and Celis, J., 1981a, ^{35}S-methionine labeled polypeptides from HeLa cells coordinates and percentages of some major polypeptides, *Cell Biol. Int. Rep.* 5:93–96.

Bravo, R., Fey, S. J., Small, J. V., Larsen, P. M., and Celis, J. E., 1981b, Coexistence of three major isoactins in a single Sarcoma 180 cell, *Cell* 25:195–202.

Chafouleas, J. G., Pardue, R. L., Brinkley, B. R., Dedman, J. R., and Means, A. R., 1981, Regulation of intracellular levels of calmodulin and tubulin in normal and transformed cells, *Proc. Natl. Acad. Sci. USA* 78(2):996–1000.

Coligan, J. E., Gates, F. T., III, Kimball, E. S., and Maloy, W. L., 1983, Radiochemical sequence analysis of biosynthetically labeled proteins, *Methods Enzymol.* 91:413–434.

Cooper, T. G., 1977, Radiochemistry, in: *The Tools of Biochemistry*, pp. 113–135, John Wiley & Sons, New York.

Duncan, R., and McConkey, E., 1982, How many proteins are there in a typical mammalian cell?, *Clin Chem.* 28(4):749–755.

Ercolani, L., Brown, T. J., and Ginsberg, B. H., 1984, Tunicamycin blocks the emergence and maintenance of insulin receptors on mitogen-activated human T lymphocytes, *Metabolism* 33:309–316.

Garrels, J. I., 1979, Two-dimensional gel electrophoresis and computer analysis of proteins synthesized by cloned cell lines, *J. Biol. Chem.* 254:1971–1977.

Geetha-Habib, M., and Bouck, G. B. 1982, Synthesis and mobilization of flagellar glycoproteins during regeneration in *Euglena*, *J. Cell Biol.* 93(2):432–441.

Hames, B. D., and Rickwood, D., 1981, Appendix II, in: *Gel Electrophoresis of Proteins, a Practical Approach*, pp. 265–277, IRL Press, Oxford, England.

Honza, R., Jork, R., Lossner, B., and Matthies, H., 1983, The effect of dopamine on L-fucose incorporation into hippocampal and striatal slices of dopamine-supersensitive rats, *Biomed. Biochim. Acta* 42(4):419–422.

Ivarie, R. D., and O'Farrell, P. H., 1978, The glucocorticoid domain: Steroid mediated changes in the rate of synthesis of rat hepatoma proteins, *Cell* 13:41–55.

Liu, T. C., and Jackson, G. L., 1985, Synthesis and release of luteinizing hormone *in vitro* by rat anterior pituitary cells: Effects of gallopamil hydrochloride (D600) and pimozide, *Endocrinology* **117**:1608–1614.

Nolan, T. J., and Farrell, J. P., 1985, Inhibition of *in vivo* and *in vitro* infertility of Leishmanta donovani by tunicamycin, *Mol. Biochem. Parisitol.* **16**(2):127–135.

O'Farrell, P. H., 1975, High-resolution two-dimensional electrophoresis of proteins, *J. Biol. Chem.* **250**:4007–4021.

Otto, A. M., and Muhlradt, P. F., 1980, Cell cycle dependent rate of labeling of cellular and secreted glycosaminoglycans in mouse embryonic fibroblasts, *J. Supramol. Struct.* **13**:281–294.

Ramagli, L. S., and Rodriguez, L. V., 1985, Quantitation of microgram amounts of protein in two-dimensional polyacrylamide gel electrophoresis sample buffer, *Electrophoresis* **6**:559–563.

Savage, K. E., and Baur, P. S., 1983, Effect of tunicamycin, an inhibitor of protein glycosylation on division of tumour cells *in vitro, J. Cell Sci.* **64**:295–306.

Schimke, R. T., 1975, Methods for analysis of enzyme synthesis and degradation in animal tissues, *Methods Enzymol.* **40**:241–266.

Taylor, T., and Weintraub, B. D., 1985, Thyrotropin (TSH)-releasing hormone regulation of TSH subunit biosynthesis and glycosylation in normal and hypothyroid rat pituitaries, *Endocrinology* **116**(5):1968–1976.

Use of Autoradiography in Polyacrylamide Gel Electrophoresis

I. INTRODUCTION

Autoradiography is a photographic method of recording the spatial distribution of radioisotopes within a particular tissue, cell, cell organelle, or molecule. In this technique, a radioactively labeled specimen, such as a tissue slice or polyacrylamide gel, is placed in direct contact with a photographic emulsion designed for autoradiography. The radioactive atoms will decay in the sample, and emitted radiation will activate individual silver halide grains in the emulsion, rendering them susceptible to conversion into metallic silver by a photographic developer.

II. ISOTOPES COMMONLY USED IN AUTORADIOGRAPHY

The most commonly used radioisotopes used for autoradiography are of three energy types: high, medium, and low (see Table 7.1). Because of the different energy types of these isotopes, the strategy for achieving optimal conditions for autoradiography will vary. It is therefore necessary to understand the half-life, emission, and energy characteristics of the isotope to be used in order to choose the type of x-ray film, as well as temperature and time of exposure conditions.

Most of the radioisotopes commonly used for this method emit β particles. These particles are electrons that are themselves easily scattered by orbital electrons. When a β particle collides with other electrons, it rapidly loses energy and is sharply deflected in each collision. Because the magnitude of this deflection is dependent on the energy of the particle, the path of the particle will vary accordingly.

Table 7.1. Radioisotopes Commonly Used in Autoradiography[a]

Isotope	Radiation	Energy (MeV)		Half-life
^{3}H	β	Low	(0.018)	12.26 years
^{14}C	β	Medium	(0.156)	5730 years
^{35}S	β	Medium	(0.167)	88 days
^{45}Ca	β	Medium	(0.256)	165 days
^{32}P	β	High	(1.71)	14.3 days
^{125}I	γ	High	(0.035)	60 days
	X		(0.027)	
	e$^-$		(0.30)	

[a]Modified from Hahn (1983) and Freifelder (1976).

III. EMULSIONS AND FILMS USED IN AUTORADIOGRAPHY

Autoradiography that employs the small specimens such as for microscopy involves the use of special liquid emulsions of stripping films. Because this is beyond the scope of this text, these are not addressed here.

The films designed for the use of larger specimens such as polyacrylamide electrophoresis gels are generally high-speed medical x-ray films. These are optimal for larger samples because they have incorporated emulsions sensitive in ability to react effectively to light (photons) as well as to ionizing radiation such as γ-ray photons, electrons and β particles. Furthermore, image-intensifying methods, such as fluorography or use of intensifying screens, can easily be employed with these films. The commonly used x-ray films consist of (1) a radiation-sensitive emulsion layer, and (2) a flexible film support. The emulsion layer is made of minute crystals of silver halide, bromides, chlorides, or mixed halides, including iodides, suspended in gelatin. Gelatin is an excellent dispersing and protective medium for silver halide grains that permits rapid penetration of the processing solutions as well as providing a stable matrix before and after processing. This radiosensitive emulsion is sandwiched between two gelatin layers, one that provides adhesion of the emulsion to the support and another that provides protection for the emulsion layer (Hahn, 1983).

IV. FORMATION OF AN AUTORADIOGRAPHIC IMAGE

As a radioactive particle passes through a nuclear emulsion, it will continually lose energy as it collides with nuclei and orbital electrons. The exposing energy will be absorbed in the silver halide grain with the release of electrons.

As the electrons move through the grain, they are trapped by surface or internal sensitivity specks, which then become negatively charged. These negatively charged sensitivity specks then attract the positively charged silver ions and an atom of metallic silver is formed. This begins the formation of the *latent image,* the term used to describe the invisible precursor of the invisible developed image. The interaction between the negatively and positively charged groups continues until a metallic silver speck of increasing size and stability is formed. Three to six atoms of silver are required to form a stable, latent image center that can be developed. It will require more than one light photon interaction to form the three to six atoms of silver. However, for ionizing radiation, such as β particles or x-ray or γ-ray photons, a single interaction is required. This latent image formed by exposure to light over a long time is not very stable, and the sensitivity specks may begin to lose their ability to be developed, because the decay process will reduce them to a size that is no longer stable. This decay can be accelerated by the presence of oxygen or moisture in the emulsion as well as by thermal agitation of the crystal lattice at room temperature. This is why the formation of the latent image can be enhanced by low-temperature exposure or preflashing (Hahn, 1983).

V. EXPOSURE METHODS FOR AUTORADIOGRAPHY

The type of exposure as well as the temperature and time of exposure will depend on the radioisotope used as well as on the specific activity of the radioisotope incorporated into the specimen (e.g., protein in poly-

Table 7.2. Recommended Conditions for Optimal Sensitivity of Exposure Using Different Radioisotopes for Autoradiography

Radioisotope	Exposure	Temperature	Type x-ray film
^3H	Fluorographic	$-70°C$	High sensitivity to UV and blue spectrum
^{35}S, ^{14}C	Fluorographic	$-70°C$	High sensitivity to UV and blue spectrum
^{35}S, ^{14}C	Direct exposure	$20°C$	High speed
^{32}P	Direct exposure	$20°C$	High speed
^{32}P	With intensifying screen	$-70°C$	High sensitivity to UV and blue spectrum
^{125}I	Direct exposure	$20°C$	High speed
^{125}I	With intensifying screen	$-70°C$	High sensitivity to UV and blue spectrum

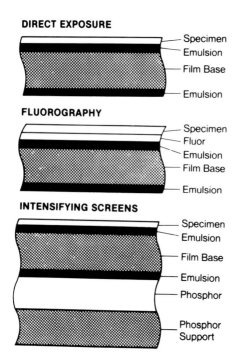

Figure 7.1. Diagram of three types of methods used for autoradiography. (From Hahn, 1983.)

acrylamide gel). The recommended exposure conditions for use with different isotopes is given in Table 7.2.

Three types of exposure methods can be used with polyacrylamide gels: (1) direct exposure; (2) fluorography; and (3) the use of intensifying screens (Fig. 7.1). The method of choice will depend on the penetration levels of the isotope used (Fig. 7.2).

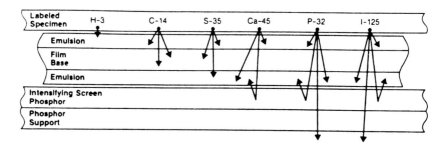

Figure 7.2. Penetration levels of isotopes commonly used for autoradiography. (From Hahn, 1983.)

A. Direct Exposure Method

With this method, the specimen is placed directly against the emulsion. Direct exposure will provide a high image quality and can be used with all β-emitting isotopes—although less optimally with tritium. This exposure will require a longer exposure time as compared with intensification or fluorography methods.

B. Fluorography Method

This method is a type of intensification procedure that involves the overlayering or the impregnation of the sample (i.e., polyacrylamide gel) with a fluor such as PPO (2,5-diphenyloxazole) or sodium salicylate. The use of an organic scintillator such as PPO will cause the conversion of the energy of the β particle to *visible light,* which will cause the exposure of a blue-sensitive x-ray film. The adsorption of β particles within the sample is therefore overcome by the increased penetration of light. The increase in the efficiency of detection of ^{14}C and ^{35}S in polyacrylamide gels can be as great as 10-fold above levels obtained with direct autoradiography (Bonner and Laskey, 1974). An isotope used that can also be effective in direct exposure methods may cause additional exposure of the film. Fluorography can therefore be used effectively to improve the detection of a radiolabeled signal.

While these methods are useful in enhancing detection methods, some concerns remain concerning effects on quantitation of images. The quantitative relationship between the absorbance of the fluorographic film image and the distribution of radioactivity in the sample is reportedly complex and is not linear with the amount of radioactivity or with time (Laskey and Mills, 1975). These studies demonstrated that the fluorographic image absorbance is dependent not only on the number of disintegrations, but also on the rate at which they occur. While this phenomenon occurs for fluorography, it was not found to occur for direct autoradiography when the isotope ^{14}C was used.

C. Intensifying Screens

Screens are now commercially available that have been designed to improve the detection of high-energy β emitters such as ^{45}Ca, ^{32}P, and γ-ray emitters such as ^{125}I. Exposure should be designed so that the double-emulsion x-ray film is sandwiched between the polyacrylamide gel and the phosphor-intensifying screen. Emissions from the radioisotope will pass through the film to produce a direct autoradiographic image. Emissions that pass completely through and beyond the film are absorbed more efficiently by the screen, where they will produce multiple photons of visible light, which are bounced back to return through the film superimposing a photographic image over the autoradiographic image.

D. Use of Hypersensitized or Preexposed Film

Preflashing x-ray film is sometimes used because it causes initial light photons to initiate latent image formation, which will later be completed by the addition of light photons from the fluorographic process. (This process is therefore ineffective when used with direct exposures to ionizing radiation.) This procedure results in the formation of a higher density with the same exposure time or the same net density with a shorter exposure time than with film that is not preflashed. It further reduces background fog, which may occur around image areas (Hahn, 1983).

E. Exposure of X-ray Film at Low Temperature

Low-temperature exposures between $-70°$ and $-80°C$ are used to stabilize latent-image formation during long exposures to visible light, which is generated during fluorographic procedures or by intensifying screens. This stabilization can result in increase in sensitivity of detection as much as fourfold or more (Hahn, 1983). It is important to note that there is no improvement of latent image formation of direct exposures of film to ionizing radiation.

REFERENCES

Bonner, W. M., and Laskey, R. A., 1974, A film detection method for tritium-labeled proteins and nucleic acids in polyacrylamide gels, *Eur. J. Biochem.* **46**:83–88.

Freifelder, D., 1976, *Physical Biochemistry: Applications to Biochemistry and Molecular Biology*, W. H. Freeman, San Francisco.

Hahn, E. J., 1983, Autoradiography: A review of basic principles. *Am. Lab.* **15**:64–71.

Laskey, R. A., and Mills, A. D., 1975, Quantitative film detection of ^3H and ^{14}C in polyacrylamide gels by fluorography, *Eur. J. Biochem.* **56**:335–341.

Chapter 8

Strategies for Use of Polyacrylamide Gel Electrophoresis in the Preparation and Characterization of Antibodies

I. INTRODUCTION

Before any investigator decides to make antibodies that can be used as probes in experiments for either experimental or clinical research, it is imperative that they decide which type of antibody is best suited for their purposes and that they plan their strategy in advance. This should guarantee a cost-effective and efficient use of antibodies. In order to make this decision, a basic understanding of the structural properties of antibodies as well as of their interactions with specific antigenic determinants is necessary.

When an animal is immunized with a foreign molecule, the *immunogen,* the animal generally responds in a variety of ways. *Immunoglobulins* (antibodies) are a heterogeneous group of serum proteins having the basic structure shown in Figure 8.1. These immunoglobulins are produced by β-lymphocytes, which combine with a region of the foreign molecule termed the *antigen.* The precise region of the antigen that is recognized by the antibody is called the *antigenic determinant,* or *epitope.* There are five major classes of immunoglobulins, IgG, IgM, IgA, IgD, and IgE, composed of two heavy peptide chains that differentiate these classes and two light chains termed κ and λ for all subclasses. The circulating IgM antibody consists of five IgM monomers connected by a joining (J) chain. A summary of the properties of human immunoglobulins is given in Table 8.1. The IgG molecule is frequently separated into proteolytic fragments in order to study the antigen-binding fragments (Fab) as compared with the complement-binding (Fc) portion (Fig. 8.2).

For the student who is just starting to make antibodies, it is important to appreciate two facts: (1) an antigen may be composed of many antigenic

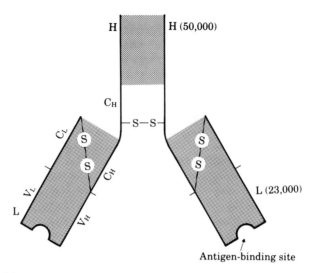

Figure 8.1. Schematic diagram of the rabbit IgG antibody molecule. (From Nisonoff, 1984.) In an individual molecule, the two H chains are identical to one another, as are the two L chains. An immunoglobulin G (IgG) molecule can contain two κ or two λ chains. Shaded areas represent regions held together by noncovalent bonds. Note that the variable regions of the L and H chains (V_L and V_H) are approximately equal in size, whereas the constant region of the H chain (C_H) is about three times as large as C_L.

determinants; and (2) a molecule that is antigenic (i.e., that is capable of being recognized by an immunoglobulin) may not necessarily be immunogenic (i.e., capable of eliciting antibodies). It is now apparent that antibodies can be made that react with almost any molecule, including such small molecules as steroids and nucleotides. Some molecules, *haptens,* will not induce antibody formation when inoculated by themselves but will induce antibodies if conjugated to a suitable carrier such as a large protein molecule.

It is further necessary to understand the difference between antibody *titer* and antibody *affinity.* The antibody titer will depend on the number of immunoglobulin molecules present as opposed to the antibody affinity, which pertains to the avidity with which the antibody will bind to its antigenic determinants. It is frequently desirable to have a high-titer antibody that has a low affinity, so that antibody–antigen complexes can easily be separated (e.g., as for use in immunoaffinity chromatography).

For additional details on basic principles of immunology, general texts by Benacerraf and Unanue (1979), Nisonoff (1984), and Eisen (1980) are recommended. The following sections address the strategies involved in

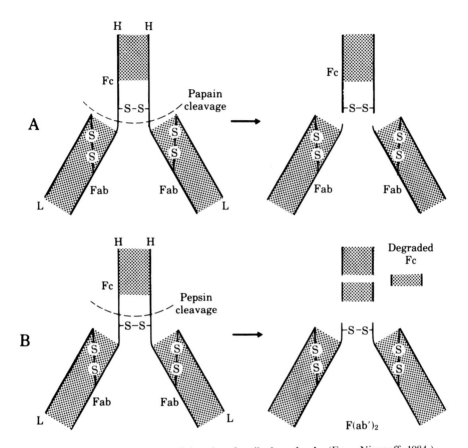

Figure 8.2. Diagram of functional domains of antibody molecule. (From Nisonoff, 1984.)

developing antibodies for use as probes that can be used for defined experimental problems.

II. STRATEGIES FOR DEVELOPING ANTIBODIES

A. Polyclonal Antibodies

When an animal is immunized with a complex immunogen such as a protein, glycoprotein, or glycolipid, many different immunoglobulins (polyclonal) are developed that collectively recognize multiple antigenic determinants. These epitopes may include structural sequences such as

Table 8.1. Summary of Properties of Human Immunoglobulins[a]

Isotype	Chain structure	H-chain designation	Mol. wt. ($\times 10^{-3}$)	No. of domains in H chain	Mol. wt. of H chain ($\times 10^{-3}$)	Percentage of carbohydrate	Normal serum conc. (mg/ml)[b]	Half-life (days)
IgG1		γ_1	146	4	51	2–3	9	21 ± 5
IgG2		γ_2	146	4	51	2–3	3	20 ± 2
IgG3		γ_3[c]	165	4	60	2–3	1	7 ± 1
IgG4		γ_4	146	4	51	2–3	0.5	21 ± 2.5
IgM		μ	970	5	72	9–12	1.2	5
IgA1		α_1	160	4	58	7–11	2.0	6
IgA2								
A$_2$m(1)		α_2	160	4	58	7–11	0.5	?
A$_2$m(2)		α_2	160	4	58	7–11	0.5	6
sIgA		α_1 or α_2	405	4	58	7–11	0–0.05	—
IgD		δ	170	4	63	10–12	0.06	3
IgE		ϵ	188–196	5	72–76	12	0.0002	2.5–4

[a]These values will vary to a certain extent among different animal species, but this summary can serve as a guideline for basic information on immunoglobulin subclasses. From Nisonoff (1984).
[b]There is considerable individual variation in these values.
[c]Ten or eleven S-S bonds link the H chains.

amino acid sequences or specific carbohydrate moieties, or they may recognize structural or conformational determinants (see Fig. 8.3). Because of these properties, there are numerous advantages for the use of polyclonal antibodies *if they are made against a highly purified protein,* such as one isolated using two-dimensional polyacrylimide gel electrophoresis (PAGE).

1. Major Advantages of Polyclonal Antibodies

a. Since multiple antigenic determinants are recognized by different immunoglobins, multiple subclass and high-affinity antibodies are likely to be present.

b. Antibodies to sequential as well as conformational antigenic determinants are likely to be developed.

c. Since the antibodies are likely to recognize multiple determinants that are specific for a protein (or other antigen), the signal-to-noise ratio obtained by most antibody-screening assays is generally greater than if one has a monoclonal antibody that may recognize a common epitope on more than one antigen. (This is extremely important if one wishes to screen a gene library using antibody-screening methods.)

d. The recognition of multiple antigenic determinants generally results in the formation of precipitating antibodies; therefore these reagents can be used with a large variety of methods.

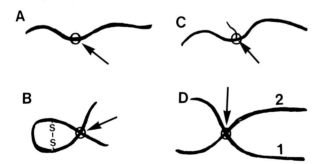

Figure 8.3. Examples of types of antigenic determinants (arrows) associated with macromolecules. (A) Sequential determinant (may be on protein or carbohydrate). (B) Antigenic determinant, dependent on secondary or tertiary structure of protein. (C) Antigenic determinant, dependent on interaction of carbohydrate side chain and protein. (D) Antigenic determinant, dependent on the interaction of multiple macromolecules (e.g., proteins 1 and 2).

2. Major Disadvantages of Polyclonal Antibodies

a. If immunogen cannot be highly purified, the chances of obtaining specific antibodies that can be used to carry out meaningful experiments are small.
b. Since multiple antigenic determinants are recognized, it is difficult to use these as probes to "dissect" the functional domains of molecules, as is possible with monoclonal antibodies.
c. The quantities of sera are generally limited, and the chance of getting identical populations of antibodies every time a single animal is bled or different animals are immunized is small. Because the subclasses as well as the antibody affinity may change throughout the immunization procedure, some variations in antibodies may be expected.

B. Monoclonal Antibodies

When an antibody-producing B lymphocyte cell is fused with a myeloma tumor cell, an immortal cell line (hybridoma) can be established that secretes a specific antibody (referred to as a monoclonal antibody) characteristic of one that the B cell produced. This method was first published by Köhler and Milstein (1975). Since that time, there have been volumes of literature produced describing the use of this technique, including a variety of excellent methods manuals for carrying out cell fusions (see reviews by Kennett *et al.*, 1980; Edwards, 1981). Since this methodology is beyond the scope of this text, we recommend these general references, which are excellent in this area. Since much of the power of this technique comes from the strategies used to immunize, screen, as well as characterize these antibodies, we have concentrated on these aspects as well as the power of the combined use of antibodies and gel electrophoresis in basic and clinical investigation. Again, before one initiates studies to develop monoclonal antibodies, it is important to determine whether this is the best way to proceed. The following guidelines should help you in making this decision.

1. Major Advantages of Monoclonal Antibodies

a. A single homogeneous antibody to a defined epitope is obtained if the proper strain of myeloma cell (one that does not secrete its own heavy and light immunoglobulin chains) is used for fusion with the spleen cells from your immunized animal. (See review texts by Kennett *et al.*, 1980, and Hurrell, 1981, for details on standardized methods for choosing cell lines.)

b. A specific antibody can be used to dissect a functional domain of a molecule.

c. An immortal cell line can theoretically be developed to produce a limitless supply of homogeneous antibody.

d. It may be possible to develop an antibody to an antigen that cannot be purified practically by conventional methods (i.e., it is feasible to use an impure antigen to gain a pure antibody).

e. Monoclonal antibodies can be used to study the biochemical and biological properties of homogeneous antibodies of a given specificity.

f. These antibodies provide a unique method to study major histocompatibility antigens.

g. "Switch" antibodies can be selected for low-affinity binding during screening for choosing antibodies that will work optimally for immunoaffinity chromatography.

h. Immunization strategies can be used so that antibody cell lines can be selected for (e.g., antibodies recognizing shared determinants in similar proteins can be obtained). For example, mice can be immunized with a protein from one species and can later be given a boost immunization with the protein obtained from a second species. The major antibody-producing cells that will be recruited from memory cells will therefore be those that recognize similar determinants on these two different proteins.

2. Major Disadvantages of Monoclonal Antibodies

a. This procedure generally takes a great deal of manpower in order to maintain cells and carry out screening procedures.

b. The tissue-culture methods necessary are generally expensive and require the use of a well-equipped facility with sterile hoods and incubators.

c. The epitope recognized by the hybridoma-produced antibody may be shared among many different antigens (this could be to a common amino acid sequence). The use of such an antibody could prove to be a nightmare if one is using this as a single probe to screen a gene library.

d. The hybridoma cell lines are frequently unstable and may be lost due to chromosomal rearrangement.

   e. Cell lines may be lost due to culture contamination, such as with
Mycoplasma or bacteria.

3. Strategies for Screening and Selecting Monoclonal Antibodies

In summary, it is critical that a great deal of consideration be taken
in order to determine the best route for developing antibodies to be used
in your experiment. Although the development of technology for the production of monoclonal antibodies has had an enormous impact on the use
of antibodies as probes in experimental as well as in clinical investigations,
the past few years have also resulted in considerable frustration for many
investigators who have retrospectively found that monoclonal antibodies
are not the panacea for all studies. The recent improvements in technologies of high-resolution two-dimensional PAGE and protein detection now
provide alternatives for the development of highly specific polyclonal
antibodies that can be used in combination with, or in place of, monoclonal
antibodies.

III. IMMUNIZATION PROTOCOLS

While many investigators take the methods used for the immunization
of animals for granted, the choice of animal, adjuvant, route of admin-
istration, and dosage, as well as the timing of boost immunizations are
extremely important. This chapter summarizes some of the conditions we
have used routinely with excellent results.

A. Choice of Antibody

The choice of animal to be used to develop antibodies will depend
on a variety of factors:

   1. *Source of immunogen:* If one wants to obtain an optimal immune
response, it is generally advisable to use an animal that is phy-
logenetically removed from the source of the immunogen. For
example, if you wish to develop a high-titer polyclonal antibody
to a specific mouse protein, it may be advisable to immunize a
rabbit rather than a rat. For example, in these studies, you would
immunize an animal with an immunogen produced by that animal

(e.g., sperm antigens into a male rabbit that produced the sperm: autoimmunization) or from different animals of the same species but of a different strain (e.g., blood cell proteins from a mouse of one strain into a mouse of a different strain: alloimmunization).
2. *Need for polyclonal vs. monoclonal antibodies:* It is imperative that an investigator decide *a priori* which type of antibody will best suit his or her purposes. If monoclonal antibodies are the antibody of choice, it is advisable to use mice or rats for immunization, as the hybridomas generated from available mice and rat cell lines tend to be more stable.

If polyclonal antibodies are preferred, it is advisable to immunize an animal from which adequate quantities of sera can be easily obtained. Now that there are a number of large commercial animal farms to which you can send your antigen for antibody preparation, it is feasible and financially practical to prepare large quantities from sheep or goats.

B. Guidelines for Choosing Animal Species to Be Immunized

The following guidelines can be used to determine which species is best suited to your needs.

1. Rabbits

a. *Major Advantages*

1. Rabbits are easy to maintain.
2. They are easily bled and can be given reasonable quantities of serum (approximately 50 ml of blood can be collected per week from a large rabbit).
3. Multiple subclasses of IgG will bind *Staphylococcus aureus* cells as well as the purified protein from this bacteria (protein A).
4. A variety of immunoglobulins directed against rabbit IgGs as well as antibody detection kits are commercially available.

b. *Major Disadvantages*

1. Quantities of serum may be limiting if large volumes of serum are needed for immunoaffinity chromatography.
2. Many laboratory rabbits are infected with diseases such as pasteurellosis, which may affect the immune response directed against your immunogen. (This can be overcome by purchasing pasteurella-free rabbits and maintaining them in pathogen-free facilities.)

2. Guinea Pigs

   a. *Major Advantages*

   1. Guinea pigs are easy to maintain.
   2. Multiple subclasses of IgG will bind *Staphylococcus aureus* cells as well as the purified protein (protein A).
   3. A variety of immunoglobulins directed against rabbit IgGs as well as antibody detection kits are commercially available.
   4. They generally make high-titer antibodies.

   b. *Major Disadvantages*

   1. Only small amounts of sera can be obtained (5–10 ml blood/bleeding).
   2. Blood must be collected using cardiac puncture methods, so there is frequently a risk of "losing" the immunized animal.

3. Mice and Rats

   a. *Major Advantages*

   1. Mice and rats can easily be used for production of monoclonal antibodies.
   2. Myeloma cell lines have been developed for these species.

   b. *Major Disadvantages*

   1. These animals are not a practical species for polyclonal antibody.
   2. Very small amounts of sera can be obtained; therefore, all subclasses of IgG do not bind protein A.

4. Sheep and Goats

   a. *Major Advantages*

   1. Large quantities of sera can be easily obtained.
   2. Immunogen can be shipped to a variety of commercial farms for custom antibody production (e.g., Bethyl Laboratories, Cappell Laboratories).

   b. *Major Disadvantages*

   1. Special facilities are needed to handle large animals, or immunogen must be sent to a commercial producer to obtain antibodies.
   2. IgG does not bind purified protein A with high affinity; therefore, second antibody methods have to be employed in some antibody detection.

C. Choice of Adjuvant

The term *adjuvant* is broadly applied to any reagent that will increase the immunological response. While a number of substances, including alum, dextran sulfate, and large polymeric anions, have been used as adjuvants, the most common and most routinely effective adjuvants are the water-in-oil emulsions developed by Freund (see discussions by Benacerraf and Unanue, 1979; Nisonoff, 1984). The adjuvant in which living or dead tubercle bacillus is suspended is referred to as *complete Freund's adjuvant,* while the water-in-oil emulsion alone is referred to as *incomplete Freund's adjuvant.* The complete Freund's adjuvant has a dual role in stimulating the immune response in that the immunogen is slowly released from the oil and the complex immunogens of the bacteria bring about an enhanced cellular response. Such adjuvants facilitate the maintenance of low effective antigen levels in tissues and provoke inflammation. As a result, macrophages can accumulate and bind antigen for reaction with B and T lymphocytes for an enhanced immunoreaction.

When a protein is suspended in a polyacrylamide gel matrix in adjuvant, this matrix can act as an additional carrier and will prolong the exposure of the immunogen to the immune system. (We have found that if a protein is eluted from the polyacrylamide gel, it is not as immunogenic as it is when not eluted.)

D. Preparation of Polyacrylamide Gel for Immunization

The most efficient and effective method for developing an antibody to a spot identified in as polyacrylamide gel is simply to cut the spot from the gel. A simple method has been developed to reduce the chances that immunogen will be lost during emulsification with adjuvant prior to immunization. This method is described in detail in Appendix 11.

E. Routes of Administration

The most common methods used for immunization include intradermal, subcutaneous, intramuscular, intravenous, subscapular, and intralinguinal (see review by Hurn and Chantler, 1980). Whereas injection of footpads of rabbits has been a frequently used procedure, this method results in a great deal of pain to the animals and will result in infection, frequently leading to the death of the animal. Since this procedure does not result in a significant increase in antibody production over other methods, we strongly recommend that this procedure not be used. Furthermore, the veterinarians in care of most animal facilities will no longer allow this procedure to be used. The recommended procedure is similar to that described by Vaitukaitis *et al.* (1971) and is one we have found to be the best method for immunization of rabbits and guinea pigs (Dunbar and Raynor, 1980; Wood and Dunbar, 1981). A similar procedure can be used for sheep and goats. To date, this is the most effective method we have found for the primary immunization with protein in polyacrylamide gels (Wood and Dunbar, 1981; Dunbar and Raynor, 1980). Exact details are given in Appendix 11.

F. Dosages for Immunization

It is very difficult to generalize the precise amount of protein or other molecules that may elicit an adequate immune response to produce antibodies. In general, the further removed the species from which your protein is isolated from the animal species to be inoculated, the better the chance that it will be immunogenic. If a protein is highly conserved, it will most likely not be very immunogenic, and you may have to conjugate it or chemically modify it in order to make it more immunogenic. It is also likely that there are serum antibodies that are not easily detected with assay methods requiring high antibody levels, such as immunoprecipitation methods. If you have an abundant protein, you may be able to concentrate and purify the antibodies by affinity chromatography so they can be used in other assays (see Appendix 11).

We have used as little as 10 μg protein (the initial immunization with 10 μg in each of two booster immunizations) in rabbits and guinea pigs to generate antibodies to proteins isolated from polyacrylamide gels. Better responses have been produced, however, when we have used 50–300 μg protein per injection. We have had antibodies made in sheep using similar amounts of proteins. Finally, we have developed monoclonal antibodies from cells of spleens of mice immunized with as little as 10 ng protein given in each of three subcutaneous injections stained with the silver stain and isolated from polyacrylamide gels.

G. Enhancement of Immunogenicity of Proteins

One of the advantages of using proteins in polyacrylamide gels for immunization is that the protein is trapped in the acrylamide matrix and is not rapidly degraded. The animal is therefore exposed to the immunogen for a long period of time. Furthermore, since large particles are generally more immunogenic, the acrylamide particles aid in the immune response.

A possible disadvantage of this method is that a detergent-denatured protein is used for immunization. The denaturation process may alter the antigenic determinants associated with the native protein and may also reduce the immunogenicity of the protein. Another advantage to this procedure, however, is that the antibodies developed should always react in immunoblotting procedures because they were developed against a denatured form of the protein and will therefore recognize epitopes on the denatured protein.

An alternative procedure for this latter problem is to modify proteins that have been electroeluted from polyacrylamide gels in order to enhance their immunogenicity. Frequently, it is necessary to modify or conjugate proteins to a peptide to other proteins that are more immunogenic. Commonly used methods include the linkage of proteins to particles to polymerize protein or to form covalent conjugates of proteins and smaller peptides. A variety of agents that are simple and effective but that do not alter the native antigenicity of proteins have been described (Erlanger, 1980; Reichlin, 1980).

1. Conjugation of Immunogen to Carrier Protein or Synthetic Antigen

a. Carbodimide Method. One commonly used method that has been used to enhance immunogenicity is to conjugate such immunogens using the carbodimide method (Bauminger and Wilchek, 1980). Carbodimides are a group of compounds whose general formula is $R—N{=}C{=}N—R'$, where R and R' are aliphatic groups. The conjugation of two compounds using this method requires the presence of an amino and a carboxyl group. If carboxyl groups are not present in the molecule to be conjugated, they may be introduced (Steiner *et al.*, 1969). The introduction of these groups may also be used when the native functional group responsible for biological activity is coupled through these groups in a manner that would mask them as antigenic determinants (Bauminger *et al.*, 1974; Jeffcoate and Searle, 1972). In most reactions, the amino groups involved in the conjugation are the lysyl residues of the protein carriers (McGuire *et al.*, 1965; Goodfriend *et al.*, 1964) or the arganyl or ananyl residues of synthetic polypeptide carriers (Arnon and Sela, 1969). This method has been used to make antibodies against peptides, peptide or steroid hormones, prostaglandins, cyclic nucleotides, and plant hormones (Arnon and Sela, 1969; McGuire *et al.*, 1965; Goodfriend *et al.*, 1964; Koch *et al.*, 1973; Bauminger *et al.*, 1973, 1974; Levine and Van Vunakis, 1970; Steiner *et al.*, 1969). The methods used for conjugation have been outlined in detail elsewhere (Bauminger and Wilchek, 1980).

b. Glutaraldehyde Method. Glutaraldehyde has also been used with good success for coupling small peptide hormones to larger proteins for immunization procedures (Reichlin *et al.*, 1968; Frohman *et al.*, 1970). Although the chemistry of the reaction of glutaraldehyde with protein has not been clearly elucidated, it is likely that several reactions occur, giving rise to a number of products. Reichlin (1980) demonstrated that this reaction gives rise to products that are stable to acid hydrolysis. He has further described detailed methods for peptide–protein conjugation using glutaraldehyde solutions.

2. Preparation of Protein Polymers for Immunization

The polymerization of many small proteins such as cytochrome C (Reichlin *et al.*, 1970) can considerably enhance their immunogenicity. The use of glutaraldehyde has been common for such polymerization procedures and the methods used are simple.

H. Immunization Procedures for Obtaining Ascites Fluid

While methods are now being routinely used to obtain monoclonal antibodies from mice that have been given hybridomas, several procedures have also been used to prepare polyclonal antibodies in large animals such as guinea pigs. Although reports of high titer antibodies have been made, this is a time-consuming procedure, and many animals may be lost before optimal antibodies can be obtained. In our experience, it has been best to obtain polyclonal antibodies from animals such as rabbits or goats, which can be easily immunized, bled, and maintained without the risk of losing the animal.

IV. COLLECTION OF BLOOD AND ASCITES FLUID FOR ANTIBODY PREPARATION

A. Bleeding Procedures

1. Mice

Small amounts of blood can be collected from mice by cutting off the tip of the tail or by bleeding from the orbital of the eye using a Pasteur pipette. Because such small amounts of blood can be obtained, mice are usually used for monoclonal antibody production using the spleens of immunized mice or for the production of antibodies in ascites fluid.

2. Rats

Blood can be collected from rats via the tail vein. This is carried out by anesthetizing the animal and then placing its tail in warm water until the vein has swollen so that blood can be drawn using a 20- to 26-gauge needle. Again, only small quantities of blood can be obtained from rats so they are generally used primarily for monoclonal antibody production.

3. Guinea Pigs

Blood can only be effectively obtained from guinea pigs using cardiac puncture on anesthetized animals. Because both anesthesia and cardiac punctate methods require a great deal of skill, it is recommended that these methods be carried out only under the supervision of a skilled investigator.

4. Rabbits

Rabbits are the most commonly used laboratory animal for in-house antibody production, and large volumes of blood can be easily obtained; these methods have been outlined in detail in Appendix 11.

5. Sheep and Goats

Large volumes of blood are obtained routinely from both sheep and goats. It is not generally cost effective to maintain these large animals in most research laboratories. Because a number of commercial sources (Bethyl Laboratories; Cappel Laboratories) are now available for preparation of custom antibodies from large animals, these methods have not been included in the text.

B. Collection of Ascites Fluid

Ascites fluid can be collected from animals that have been anesthetized or that have been euthanized. For mice, we have found it most efficient to kill mice by cervical dislocation at a specific time following injection with myeloma cells to collect as much ascites as possible rather than attempt to collect on multiple days. This time period is usually 10–20 days after injection of cells into mice that have been primed for 7 days with an irritant such as pristane (Zola and Brooks, 1981). If anesthetized animals are to be used, a syringe with a 20-gauge needle can be used. An investigator or technician should be skilled at this procedure, since the needle might puncture vital organs. If the animal is killed prior to ascites fluid collection, a Pasteur pipette can be used to collect ascites fluid both rapidly and efficiently.

V. IMMUNOGLOBULIN PURIFICATION

A. Immunoglobulin Fractionation

Although it is frequently adequate to use unfractionated serum for some studies such as with immunoblotting techniques, it is sometimes necessary to use a more purified immunoglobulin fraction. While most antibodies will not be substantially altered by most procedures, the investigator should be aware that the properties (e.g., affinity) of some antibodies may be markedly altered by some procedures such as low or high pH elution used during affinity chromatography.

1. Ammonium Sulfate Fractionation

This method is one that has long been used routinely as a crude fractionation of serum. This procedure is outlined in detail in Appendix 11 and is used to separate the majority of hemoglobin and albumin from the serum fraction. This fraction will contain multiple subclasses of immunoglobulins as well as a variety of other serum proteins, and therefore cannot be used if purified immunoglobulins are needed.

2. DEAE Chromatography

This ion-exchange chromatography procedure is used routinely to separate immunoglobulin subclasses as well as other contaminating serum proteins. Because this ion-exchange resin is commercially available in preswollen form, this is a straightforward procedure that can readily be used to purify large quantities of immunoglobulins. A typical DEAE profile of fractionated immunoglobulins is illustrated in Figure 8.4.

3. Protein A Chromatography

Investigations into staphylococcal proteins have demonstrated that a protein fraction, termed protein A (SpA), exhibited a unique reactivity with immunoglobulins (Forsgren and Sjoquist, 1966; see also review by Langone, 1982b). It was concluded from these studies that SpA reactivity

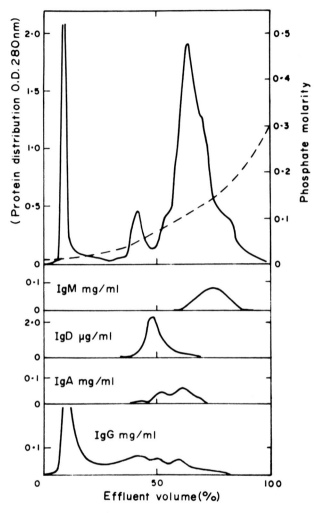

Figure 8.4. Profile of immunoglobulin subclasses obtained using DEAE chromatography. (From Fahey and Terry, 1978.)

is directed to the Fc region of IgG, and this reaction was classified as a nonspecific pseudoimmune reaction as opposed to the specific binding of antigen at antibody Fab sites. Since these studies, it has been reported that sera of all 65 mammalian species representing 7 classes and 30 orders of vertebrates react with protein A by either precipitating this molecule or by competing for binding to human myeloma IgG (Kronvall *et al.*, 1970,

1974). Because of the different types of reactivity displayed by different globulins of different species, it is important to appreciate the variations in types of binding. It is also important to realize that the affinity of binding by immunoglobulins may be very different between protein A when it is on the surface of the intact bacteria and when it is isolated and purified. For example, we have successfully used formaldehyde-fixed *Staphylococcus aureus* cells (Pansorbin, Calbiochem Chemical Co.) to precipitate goat immunoglobulin but have found that purified protein A does not bind to the same immunoglobulin in our other assays, such as immunoblots. We therefore use a second antibody, rabbit anti-goat immunoglobulin, before using [^{125}I]protein A as the probe.

Although the reactions between protein A and IgG have been most extensively studied, additional studies have shown that other subclasses of immunoglobulins are also reactive (Harboe and Folling, 1974). Furthermore, binding sites may reside in the Fab region of the IgG molecule (see Fig. 8.2) as well as on the Fc portion of the IgG molecule (Inganas *et al.*, 1980; Inganas, 1981). The complexities of these interactions have been reviewed in detail by Langone (1982*a,b*).

In summary, it is clear that while protein A has been and will continue to be a valuable probe for antibody detection, it is once again important that an investigator not overinterpret his or her data by making the assumption that this molecule simply binds to the Fc region of IgG subclasses of all immunoglobulins. It is further necessary to determine whether this probe is the best to be used for the species or subclass of immunoglobulin. In general, we have used protein A as a probe with excellent results with sera from rabbit, human, guinea pig, and nonhuman primates. For sheep, goat, or mouse sera, we routinely use a second antibody (rabbit anti-sheep, -goat, or -mouse IgG) if we plan to use protein A as the probe. Alternatively, we use the avidin–biotin-conjugated antibody systems for antibody detection, as outlined in Appendix 12.

The use of protein A Sepharose (Pharmacia) for IgG purification has become a popular method for IgG purification over the past few years. While this can be an easy, efficient method for rapidly purifying small quantities of IgG (e.g., to remove nucleases from serum samples so antibody can be used in studies such as polysome immunoprecipitation), there are limitations to this method: (1) not all classes of immunoglobulins bind protein A, and (2) there is marked species variation in immunoglobulin subclass binding to this molecule among animal species. A further complication is that the immunoglobulins of some species will bind protein A in its native form when associated with the initial bacteria but will not bind efficiently to the purified form of protein A (Langone, 1982*b*). Depending on the source of your antibody, you will have to determine whether this procedure is optimal for immunoglobulin purification.

4. Affinity Chromatography Purification

Where possible, this method is the best for purification of specific antibodies to a defined antigen. Polyclonal immunoglobulin fractions always contain antibodies that may recognize other determinants such as those on bacteria to which the animal has been exposed through the natural environment. Furthermore, because adjuvants such as complete Freund's adjuvant contain bacteria, the immunization procedure itself will induce a variety of nonspecific antibodies. If one wishes to use a polyclonal antibody to screen a gene library that uses bacteria, this can be a significant problem. The use of affinity chromatography to purify specific antibodies will help guarantee that antibodies will be specific for the expressed gene product of interest and may save considerable time in the long run.

B. Immunoglobulin Characterization by One- and Two-Dimensional Polyacrylamide Gel Electrophoresis

Immunoglobulins are a very heterogeneous population of proteins. This protein heterogeneity can be attributed to multiple protein species produced biosynthetically as products of distinct structural genes or due to posttranslational modification (i.e., glycosylation of the heavy-chain protein component) (Williamson, 1978). For many years, immunoglobulins have been characterized by electrophoretic methods. Commonly used analysis includes isoelectric focusing (IEF) of immunoglobulins (see review by Williamson, 1978) and one-dimensional sodium dodecyl sulfate–polyacrylamide gel electrophoresis (SDS–PAGE), as illustrated in Figure 8.5. Some of the IEF procedures have been established to analyze the relative charges of the entire immunoglobulin molecule, while one-dimensional SDS–PAGE separates the molecule into its constituent peptides (i.e., heavy versus light chains). High-resolution two-dimensional PAGE can now be used to separate and characterize rapidly both polyclonal as well as monoclonal antibodies. Examples of this type of characterization are shown in Figure 8.6.

VI. METHODS FOR ANTIBODY DETECTION

A. Introduction

As with any other type of experimental method, it is important to determine which antibody detection method is best suited for your needs. The following guidelines should be helpful in determining which type of method should be used as well as in the interpretation of the results.

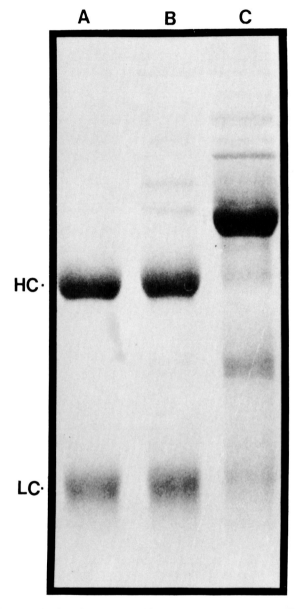

Figure 8.5. Characterization of immunoglobulins analyzed by one-dimensional sodium dodecyl sulfate–polyacrylamide gel electrophoresis (SDS–PAGE) and Coomassie Blue staining (10% acrylamide). (A) Immunoglobulins purified by immunoaffinity chromatography. HC, heavy chains; LC, light chains. (B) Immunoglobulins purified by DEAE chromatography. (C) Serum protein profile. (From S. M. Skinner and B. S. Dunbar, unpublished observations.)

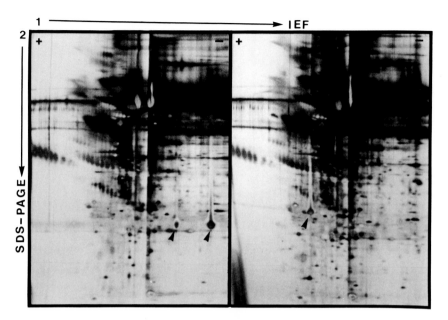

Figure 8.6. Characterization of monoclonal antibodies in tumor ascites by two-dimensional polyacrylamide gel electrophoresis. Identification of light-chain peptide species of immunoglobulins (arrows) in ascites fluid obtained from mice injected with two different hybridoma cell lines. Note the difference in relative charge of the light chains.

B. Immunoprecipitation Methods

1. Immunodiffusion and Immunoelectrophoresis

These antibody detection methods rely on the interaction of antibody with antigen to form a precipitation lattice. These reactions are frequently carried out in a matrix such as agarose, which permits optimal diffusion or electrophoretic mobility. Initially, the most commonly used form of simple diffusion was that described by Ouchterlony (1958), in which antigen is placed on one well punched in the agarose while antibody is placed in a second well (shown diagrammatically in Fig. 8.7). Alternatively, antibody is mixed within the agarose and the antigen is placed in a well and allowed to diffuse until it reaches an antigen–antibody ratio suitable for precipitation (radio immunodiffusion) (Fig. 8.8).

This is in contrast to immunoelectrophoresis methods, in which antigen is forced to migrate in an electrophoretic field into agarose containing antibody. In immunoelectrophoresis, a buffer system is chosen at the pH level at which immunoglobulins have a reduced electrophoretic mobility.

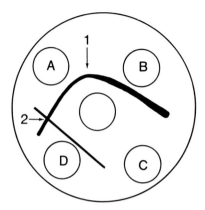

Figure 8.7. Ouchterlony double-diffusion method. In this sample, the antibody is placed in the center well and antigens are placed in the surrounding wells. This example illustrates that (1) wells A and B contain an antigen that shares immunological identity, as evidenced by the fused immunoprecipitation arc (arrow 1) (these immunoprecipitation lines may contain more than antigen–antibody complex); (2) wells A and D contain antigens that do not share immunological identity because the immunoprecipitation lines cross (arrow 2); and (3) well C does not appear to contain an antigen that is recognized by the antibody (this lack of immunoprecipitation may simply be due to an inadequate antigen : antibody ratio, so it is not possible to say conclusively that there is no antigen in well C). For a more detailed description of interpretation of complex immunodiffusion patterns, e.g., partial identity, refer to Ouchterlony and Nilsson, 1978.)

The basic types of immunoelectrophoresis methods that are commonly used are illustrated in Figures 8.9–8.14; the precise method used for carrying out immunoelectrophoresis is outlined in Appendix 11. These methods have been commonly used, and each has both advantages and disadvantages.

a. Advantages of Simple Immunodiffusion (Ouchterlony) and Radial Immunodiffusion

1. These techniques are simple to perform.
2. They require minimal equipment and are inexpensive.
3. Conformation or structural antibodies can be recognized, since proteins usually not denatured.

b. Disadvantages of Immunodiffusion

1. Precipitation reactions are dependent on precise antigen : antibody ratios; therefore, negative reactions are *never* conclusive.
2. Salts such as ammonium sulfate or detergents in samples can give precipitin reactions that can result in false-positive reactions. (These can sometimes be avoided if samples are extensively dialyzed or if agarose is thoroughly washed, dried, and stained for protein; see method in Appendix 11.)
3. The reaction usually requires significant quantities of antigen as well as antibody; therefore, it is not feasible to carry out large numbers of assays.
4. Multiple antigen : antibody precipitation reactions may occur at

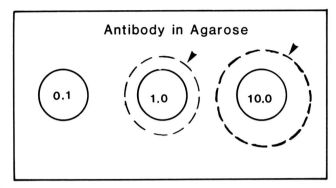

Figure 8.8. Radial immunodiffusion. Wells are cut out of agarose containing antibody. Antigens of different dilutions (e.g., 0.1–10 μg) are placed in the wells and are then allowed to diffuse into the agarose. When the antigen : antibody ratio is optimal, an immunoprecipitate (arrow) is formed. The greater the quantity of antigen, the greater the distance from the well that the immunoprecipitate ring will form.

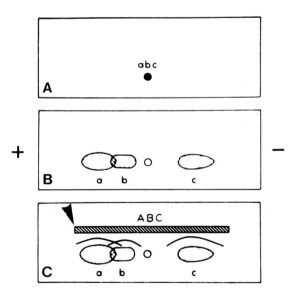

Figure 8.9. Diagram of immunoelectrophoresis. A sample containing three antigens (a,b,c) is placed in a well in agarose. Electrophoresis is carried out (A) in order to separate the antigens according to their charge (B). (C) A trough (arrow) is cut into the agarose after electrophoresis and an antibody sample (A,B,C) containing antibodies that react against antigens (a,b,c) is added and allowed to diffuse. If the proper antigen : antibody ratios are present, immunoprecipitation arcs form between the electrophoretically separated antigens and antibodies. (From Ouchterlony and Nilsson, 1978.)

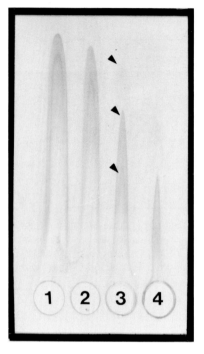

Figure 8.10. Rocket immunoelectrophoresis. Antibody is suspended in the agarose gel. Varying concentrations of antigen are placed in wells (1–4) containing increasing to decreasing concentrations of antigen, which are made in the agarose gel. Electrophoresis is then carried out using a buffer system in which immunoglobulins have minimal electrophoretic mobility. Note that the height of the "rockets" varies with the antigen concentration. The greater the concentration (well 1), the greater the distance the antigen will migrate before reaching the antigen : antibody ratio necessary for immunoprecipitation. Also note that at least three distinct rockets or antigens (arrows) can also be distinguished with this method.

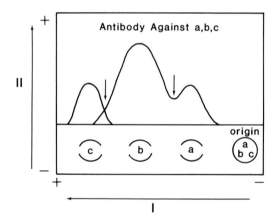

Figure 8.11. Crossed immunoelectrophoresis (two-dimensional immunoelectrophoresis). An antigen mixture (a,b,c) is separated by electrophoresis in agarose in the first dimension (I), followed by electrophoresis into agarose containing antibody against all three antigens (a,b,c). The crossed immunoprecipitation lines (arrows) demonstrate that these antigens are not similar. Note that because the antigens can be separated, it is possible to use the immunoprecipitation itself to reimmunize an animal to obtain an antibody to a highly purified antigen. For preparation of a more purified antigen, the tip of one of these precipitation areas can be cut from unstained agarose gel and can be used to immunize another animal. The antigen : antibody complex can be used directly if the same species of animal is immunized from which the original antibody was obtained.

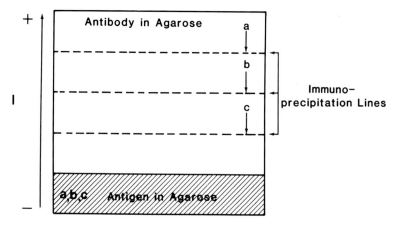

Figure 8.12. Line immunoelectrophoresis. A mixture of antigens (a,b,c) is suspended in a strip of agarose adjacent to agarose containing antibody to these antigens (a,b,c). During electrophoresis, immunoprecipitation lines form as each antigen migrates to a point at which the antigen : antibody ratio is optimal.

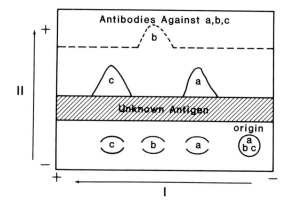

Figure 8.13. Diagram of crossed-line immunoelectrophoresis. Antigens (a,b,c) are separated electrophoretically in the first dimension (I) in agarose. Following this electrophoretic step, a strip of agarose containing an unknown antigen is placed in a strip between the antigens and agarose containing antibody against antigens a, b, and c. If the unknown antigen is recognized by the antibody, an immunoprecipitation line of antigen will be formed as the optimal antigen : antibody ratio is reached during electrophoresis in the second dimension. If one of the antigens separated in the first dimension, similar to that in the unknown antigen strip (e.g., antigen B), the ratio of antigen to antibody will be greater; therefore, the immunoprecipitation line will continue to be extended in that region of the agarose. Antigens that are dissimilar to that in the unknown antigen strip will begin to form immunoprecipitation peaks at the origin of the antibody–agarose gel. (*Note:* in all experiments, the optimal antibody : antigen ratio for immunoprecipitation of one or more antigens must be determined first in order to obtain meaningful results.)

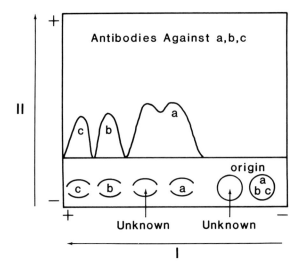

Figure 8.14. Diagram of tandem-crossed immunoelectrophoresis. Electrophoresis of two different antigen samples in tandem is carried out in agarose in the first dimension (I). Electrophoresis is then carried out in the second dimension (II) into agarose containing antibody suspended in agarose. If the unknown antigen exhibits identity with one of the antigens in the sample containing a,b,c, the immunoprecipitation will fuse (arrow).

the same place in the agar, making it difficult to determine the specificity of the antibody.

5. It does not work with most monoclonal antibodies because these are generally specific, so that sufficient numbers of epitopes (antigenic determinants) are not present on an individual molecule to cause a precipitation lattice to be formed.

c. Advantages of Immunoelectrophoresis

1. Complex polyclonal antiserum as well as antigen mixtures can be analyzed in detail.
2. Because antigens are separated electrophoretically and the agarose contains antisera, different antigens are likely to reach the proper antigen : antibody ratios to ensure that multiple immunoprecipitation reactions will occur.
3. Conformational or structural antigens can be detected, since denaturing conditions do not have to be used.
4. Antibody specificity can be analyzed in more detail because multiple antigen : antibody complexes can be detected simultaneously.

5. Antigen identity or similarity in two different samples can be determined. (This is in contrast to immunoblotting methods, in which only one of many epitopes might be shared; therefore, positive immunoblots would be obtained, even though most of the epitopes would be dissimilar.) If multiple epitopes are identical, it is likely that you would obtain fused immunoprecipitation reactions, indicating that two antigens are related or identical.

6. Methods can be used to quantitate relative levels of antibody or antigen if standards are available.

7. A "dirty" antiserum can be used to purify a specific antigen. (For example, if crossed immunoelectrophoresis is carried out and distinct immunoprecipitation arcs obtained, the top of the immunoprecipitation arc can be used to immunize an animal of the same species as the original antisera; see description in Fig. 8.11.) This method can therefore be used to purify conformationally intact antigens because denaturing conditions do not have to be used during electrophoresis in agarose.

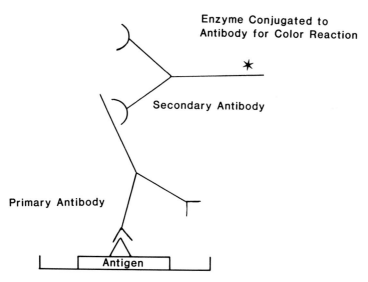

Figure 8.15. Schematic diagram of solid-phase immunoassays. Indirect enzyme-linked immunosorbent assay (ELISA) method in which antigen is immobilized on a solid surface such as a plastic microtiter well. Following blocking of nonspecific sites, antigen is incubated with primary antibody, washed and then incubated with second antibody which is conjugated to enzyme (star). After remaining unbound antibody, an enzyme substrate is used to develop a color reaction. The amount of color will be related to the amount of enzyme-conjugated antibody bound.

d. Disadvantages of Immunoelectrophoresis Methods

1. Equipment designed for immunoelectrophoresis is necessary.
2. Large quantities of antibodies (and/or antigen) are usually necessary).
3. These methods do not work with most monoclonal antibodies because these are specific for a single determinant, which usually is not sufficiently abundant for formation of an immunoprecipitation lattice to be formed.
4. You must take into account the differences in charges of antigens to ensure that all charged species will migrate into the antibody-containing agarose. (It may be necessary to carry out electrophoresis of antigen into antibody-embedded agarose, both cathode and anode.)
5. A variety of modification can be used to compare multiple antigens and antibodies directly. These methods include line, crossed-line,

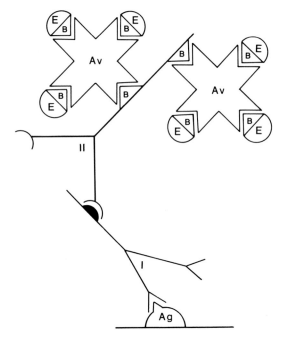

Figure 8.16. Schematic diagram of avidin-biotinylated antibody assay. This procedure is similar to that described in Figure 8.15, except that the sensitivity is enhanced by the use of the avidin–biotin complex. Ag, antigen; I, primary antibody; II, secondary antibody; Av, avidin; B, biotin; E, enzyme.

and tandem crossed immunoelectrophoresis (see Figs. 8.12–8.15). A detailed description of uses for these methods is given (Arelsen *et al.*, 1973).

C. Solid-Phase Assays

1. Enzyme-Linked Immunosorbent Assay

Solid-phase enzyme-linked immunosorbent assays (ELISA) have been used for many years (Engvall and Perlman, 1971) but have become increasingly popular, since technology has advanced for rapidly and efficiently carrying out large numbers of assays. The principle of the solid-phase enzyme-linked immunoassay is diagrammed in Figure 8.15. A va-

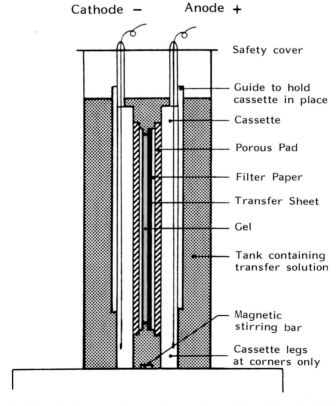

Cathode −   Anode +

Safety cover

Guide to hold cassette in place

Cassette

Porous Pad

Filter Paper

Transfer Sheet

Gel

Tank containing transfer solution

Magnetic stirring bar

Cassette legs at corners only

Figure 8.17. Schematic diagram of a typical apparatus used for electrophoretic transfer of proteins. (From Symington, 1984.)

Table 8.2. Summary of Some Major Factors Affecting Protein Blotting during Electrophoretic Transfer

Factor	Comments	References
Matrix used to immobilize protein Nitrocellulose (film of nitric acid esterified cellulose)	Advantages High binding capacity (80 $\mu g/cm^2$) Blockage of unoccupied binding sites Easily handled No preactivation required Disadvantages Binding of small proteins ($>20,000$ M_r) may be destabilized during washing Multiple probing beyond three cycles is not efficient Transfer without methanol may diminish binding	Bers and Garfin (1985) Burnette (1981) Gershoni and Palade (1983) Erickson et al. (1982)
DBM- and DPT-cellulose papers	Advantages Covalent binding, hence retention of small as well as large-molecular-weight proteins Blots can be reprobed many times Disadvantages Requires preactivation Binding dependent on pH and temperature Protein-binding capacities are generally low (25–50 $\mu g/cm^2$) Resolutions may be poor due to textured surfaces	Bers and Garfin (1985) Bittner et al. (1980) Symington (1984) Stellwag and Dahlberg (1980) Gershoni and Palade (1983)
Cationic nylon (Zeta-probe, Biorad Zeta-Bind AMF, Cuno), nylon 66 (Gene Screen, du Pont)	Advantages High binding capacity (480 $\mu g/cm^2$) Binding capacity not influenced by methanol in transfer buffer	Bers and Garfin (1985) Gershoni and Palade (1982) Van Eldik and Wolcheck (1984)

Antibody probes
[125I]Protein A
	Disadvantages
	  Blocking unoccupied binding sites difficult
	  High nonspecific binding following transfer
	Advantages
	  Easy to use
	  Commercially available
	  Binds to IgG of many animal species
	Disadvantages
	  Does not bind Ig of all species or subclasses of IgG
	  Iodinated form loses reactivity over time
		Symington (1984)
		Langone (1982b)
		Dunbar (unpublished observations)

Second antibody	Advantages
	  Affinity-purified antibodies can be obtained commercially for most animal species
	  Ig can be iodinated for direct use in blots
	  Can be used with avidin–biotin systems for enhanced sensitivity
	Disadvantages
	  Need specific second antibody (preferably affinity purified); therefore may be more expensive	

Blocking reagents
Bovine serum albumin
	Advantages
	  Partially purified; therefore, most antibodies will not cross-react
	Disadvantages
	  Expensive method if many blots are carried out
		Symington (1984)
		Bers and Garvin (1985)

Blotto (instant nonfat dry milk)	Advantages
	  Commercially available in large quantities
	  Effectively blocks most binding sites on immunoblotting matrices
	Disadvantages
	  Complex carbohydrates in milk; therefore, these absorb out antibodies that recognize carbohydrate determinants
		Johnson et al. (1984)

(Continued)

Table 8.2. Summary of Some Major Factors Affecting Protein Blotting during Electrophoretic Transfer (Continued)

Factor	Comments	References
Transfer buffers	Depends on type of gel and immobility matrix—most common in Tris–glycine, pH 8.3 with SDS (see Appendix II)	Bers and Garvin (1985) Towbin and Gordon (1984) Anderson et al. (1982)

Methanol in transfer buffer	Advantages
	  Remove SDS from protein; therefore, some proteins may bind more efficiently
	  May effect binding affinity and capacity to immobilize matrix
	  Keeps polyacrylamide gels from swelling during transfer
	Disadvantages
	  Removes SDS; therefore, charge on proteins will not be uniform and transfer may be less quantitative
	  Causes general reduction in pore size, thus resisting transfer of some proteins
	  May cause proteins to precipitate in gel
	  May alter antigenic determinants
		Bers and Garvin (1985)
		Anderson et al. (1982)
		Dunbar (unpublished observations)

riety of methods and enzyme systems have now been developed, as well as many commercially available kits, including the avidin–biotin systems, which are extremely sensitive because of the multiple binding properties and high affinity of avidin for biotin (see Fig. 8.16 and Appendix 11). Frequently, it is desirable to obtain a quick "yes or no" answer as to whether antibodies to your antigen are present. This is particularly important if you are screening large numbers of clones to detect monoclonal antibodies. We have found that the most efficient method for initial detection of antibodies is the solid-phase enzyme-linked assay. The detailed methods for one of these assays is outlined in Appendix 11. This is one of the most sensitive assays for antibody detection and does not rely on the proper ratio of antigen to antibody as do immunoprecipitation assays. Your chances of missing the presence of antibody may therefore be less than they would be with immunoprecipitation methods.

a. Major Disadvantages of ELISA Methods

1. This assay may require significant quantities of antigen to coat microtiter plates.
2. Although assays can be carried out simply in test tubes, this method usually requires special equipment (e.g., microtiter plate absorbance reader; multiple pipettor) for efficient, routine assays (see Appendix 11).
3. Conditions need to be optimized for optimal adherance of each different antigen to microtiter plates (see Appendix 11).

b. Major Advantages of ELISA Methods

1. This assay is sensitive, and high-quality reagents can be obtained commercially for standardization of your assay.
2. It is sometimes possible to use crude antigen to coat microtiter plates.
3. Proteins electroeluted from polyacrylamide gels can be used to coat microtiter plates; therefore, immunoblots can be screened for antibodies that detect denatured proteins.
4. Assays can be standardized to quantitate both antigen and antibody.
5. Large numbers of assays can be conveniently and accurately processed.

2. Immunoblot (Western Blotting) Methods

During the past few years, one of the most commonly used methods for identifying specific proteins recognized by antibodies has been the procedure that has become popularly known as Western blotting (in contrast to Southern and Northern blotting techniques, which refer to the identification of nucleic acids). These methods involve the transfer (usually electrophoretic) of proteins to an immunoblotting matrix such as nitrocellulose paper, followed by identification of molecules with a specific probe such as an antibody lectin. A diagram of a typical electrophoretic transfer setup is shown in Figure 8.17 and is pictured in Appendix 12.

In general, electrophoretic transfers can be highly controlled, but there are many parameters that affect the efficiency of transfer (see reviews by Bers and Garfin, 1985; Gerschonic and Palade, 1983; Symington, 1984; Towbin and Gordon, 1984; Tsang et al., 1985). These parameters have been exhaustively covered previously and are summarized in Table 8.2. More recently, an apparatus has been made available (from Sartorius Co.) that can be used to transfer proteins without the use of buffer solutions. This method is more cost effective because the only buffer solutions are those that are incorporated in filters between polyacrylamide gels (see Appendix 12).

a. Major Advantages of Immunoblotting Method

1. This method can directly determine which proteins are being recognized by an antibody (or other specific probes), especially if high-resolution two-dimensional PAGE is used.
2. It is extremely sensitive for antigen detection.
3. It is easily carried out because commercial reagents (e.g., [^{125}I]protein A or enzyme-conjugated second antibodies) are now available.

b. Major Disadvantages of Immunoblotting Methods

1. There may be variation in the binding of different proteins to different types of papers; this binding also depends on buffers or media used to bind antigens. Studies designed to optimize binding of your protein(s) will therefore have to be carried out. Even then, quantitation of proteins may be difficult.
2. Denatured proteins separated by polyacrylamide gels are generally used in transfers; therefore, conformational or structural antigenic determinants may be destroyed before or during transfer (particularly if alcohol is used in the transfer buffer). This method therefore cannot be used to determine definitively the specificity of an antibody.

REFERENCES

Anderson, N. L., Nance, S. L., Pearson, T. W., and Anderson, N. G., 1982, Specific antiserum staining of two-dimensional electrophoretic patterns of human plasma proteins immobilized on nitrocellulose, Electrophoresis 3:135–142.

Arnon, R., and Sela, M., 1969, Antibodies to a unique region in lysozyme provoked by a synthetic antigen conjugate, *Proc. Natl. Acad. Sci. USA* **62**:163–170.

Axelson, N. H., Kroll, J., and Weeke, B., 1973, A manual of quantitative immunoelectrophoresis: Methods and applications, *Scand. J. Immunol.* (Suppl. 1) **2**:1–169.

Bauminger, S., and Wilchek, M., 1980, The use of carbodimides in the preparation of immunizing complexes, *Methods Enzymol.* **70**:151–159.

Bauminger, S., Lindner, H. R., and Weinstein, A., 1973, Properties of antisera to progesterone and to 17-hydroxy-progesterone elicited by immunization with the steroids attached to protein through position 7, *Steroids* **21**:947–956.

Bauminger, S., Kohen, F., and Lindner, H. R., 1974, Steroids as haptens: Optimal design of antigens for the formation of antibodies to steroid hormones, *J. Steroid Biochem.* **5**:739–747.

Bauminger, S., Lindner, H. R., Perel, E., and Arnon, R., 1969, Antibodies to a phytooestrogen: Antigenicity of genistein coupled to a synthetic polypeptide, *J. Endocrinol.* **44**:567–578.

Benacerraf, B., and Unanue, E. R., 1979, *Textbook of Immunology,* Williams & Wilkins, Baltimore.

Bers, G., and Garfin, D., 1985, Protein and nucleic acid blotting and immunobiochemical detection, *BioTechniques* **3**(3):276–288.

Bittner, M., Kupferer, P., and Morris, C. F., 1980, Electrophoretic transfer of proteins and nucleic acids from slab gels to diazobenzyloxymethyl cellulose or nitrocellulose sheets, *Anal. Biochem.* **102**:459–471.

Burnette, W. N., 1981, "Western blotting": Electrophoretic transfer of proteins from sodium dodecyl sulfate polyacrylamide gels to unmodified nitrocellulose and radiographic detection with antibody and radioiodinated Protein A, *Anal. Biochem.* **112**:195–203.

David, G. S., Wang, R., Bartholomew, R., Sevier, E. D., Adams, T. H., and Greene, H. E., 1981, The hybridoma—An immunochemical laser, *Clin. Chem.* **27**:1580–1585.

Dunbar, B., and Raynor, D., 1980, Characterization of porcine zona pellucida antigens, *Biol. Reprod.* **22**:941–954.

Edwards, P. A. W., 1981, Review article: Some properties and applications of monoclonal antibodies, *Biochem. J.* **200**:1–10.

Eisen, H. N., 1980, Immunology: Introduction to immune responses, in: *Microbiology,* 3rd ed. (B. D. Davis, R. Dulbecco, H. N. Eisen, and H. S. Ginsberg, eds.), pp. 351–358, Harper & Row, Hagerstown, Maryland.

Engvall, E., 1980, Enzyme immunoassay, ELISA and EMIT, *Methods Enzymol.* **70**:419–438.

Engvall, E., and Perlman, P., 1971, Enzyme-linked immunosorbent assay (ELISA). Quantitative assay of immunoglobulin G, *Immunochemistry* **8**:871–884.

Erickson, P. F., Minier, L. N., and Lasher, R. S., 1982, Quantitative electrophoretic transfer of polypeptides from SDS polyacrylamide gels to nitrocellulose sheets: A method for their re-use in immunoautoradiographic detection of antigens, *J. Immunol. Methods* **51**:241–249.

Erlanger, B. F., 1980, The preparation of antigenic hapten-carrier conjugates: A survey, *Methods Enzymol.* **70**:85–104.

Fahey, J. L., and Terry, E. W., 1978, Ion chromatography and gel filtration, in: *Handbook of Experimental Immunology* (D. M. Weir, ed.), pp. 8.1–8.16, Blackwell Scientific Publications, Oxford, England.

Forsgren, A., and Sjoquist, J., 1966, "Protein A" from *S. aureus*. I. Pseudoimmune reaction with human gamma-globulin, *J. Immunol.* **97**:822–827.

Frohman, L. A., Reichlin, M., and Sohal, J. E., 1970, Immunologic and biologic properties of antibodies to a glucagon-serum albumin polymer, *Endocrinology* **87**:1055–1061.

Gershoni, J. M., and Palade, G. E., 1982, Electrophoretic transfer of proteins from sodium dodecyl sulfate polyacrylamide gels to a positively charged membrane filter, *Anal. Biochem.* **124:**396–405.

Gershoni, J. M., and Palade, G. E., 1983, Protein blotting: Principles and applications, *Anal. Biochem.* **131:**1–15.

Goodfriend, T. L., Levine, L., and Fasman, G. D., 1964, Antibodies to bradykinin and angiotensin: A use of carbodiimides in immunology, *Science* **144:**1344–1346.

Harboe, M., and Folling, I., 1974, Recognition of two distinct groups of human IgM and IgA based on different binding to staphylococci, *Scand. J. Immunol.* **3:**471–482.

Hurn, B. A. L., and Chantler, S. M., 1980, Production of reagent antibodies, *Methods Enzymol.* **70:**104–142.

Hurrell, J. G. R., 1981, *Monoclonal Hybridoma Antibodies: Techniques and Applications,* CRC Press, Boca Raton, Florida.

Inganas, M., 1981, Comparison of mechanisms of interaction between protein A from *Staphylococcus aureus* and human monoclonal IgG, IgA and IgM in relation to the classical FC gamma and the alternative F(ab')2 epsilon protein A interactions, *Scand. J. Immunol.* **13:**343–352.

Inganas, M., Johansson, S. G., and Bennich, H. H., 1980, Interaction of human polyclonal IgE and IgG from different species with protein A from *Staphylococcus aureus:* Demonstration of protein A-reactive sites located in the Fab' 2 fragment of human IgG, *Scand. J. Immunol.* **12:**23–31.

Jeffcoate, S. L., and Searle, J. E., 1972, Preparation of a specific antiserum to estradiol-17 coupled to protein through the B-ring, *Steroids* **19:**181–188.

Johnson, D. A., Gautsch, J. W., Sportsman, J. R., and Elder, J. H., 1984, Improved method for utilizing nonfat dry milk for analysis of proteins and nucleic acids transferred to nitrocellulose, *Gene Anal. Technol.* **1:**3–8.

Kennett, R. H., McKearn, T. J., and Bechtol, K. B., 1980, *Monoclonal Antibodies,* Plenum Press, New York.

Koch, Y., Wilchek, M., Fridkin, M., Chobseing, P., Zor, U., and Lindner, H. R., 1973, Production and characterization of an antiserum to synthetic gonadotropin-releasing hormone, *Biochem. Biophys. Res. Commun.* **55:**616–622.

Köhler, G., and Milstein, C., 1975, Continuous culture of fused cells secreting antibodies of pre-defined specificities, *Nature (Lond.)* **256:**495–497.

Kronvall, G., Seal, U. S., Finstad, J., and Williams, R. C., Jr., 1970, Phylogenetic insight into evolution of mammalian Fc fragment of gamma G globulin using staphylococcal protein A, *J. Immunol.* **104:**140–147.

Kronvall, G., Seal, U. S., Svensson, S., and Williams, R. C., Jr., 1974, Phylogenetic aspects of staphylococcal protein A-reactive serum globulins in birds and mammals, *Acta Pathol. Microbiol. Scand.* (*B*) **82:**12–18.

Langone, J. J., 1982*a,* Protein A of *Staphylococcus aureus* and related immunoglobulin receptors produced by streptococci and pneumococci, *Adv. Immunol.* **32:**157–252.

Langone, J. J., 1982*b,* Review article: Applications of immobilized protein A in immunochemical techniques, *J. Immunol. Methods* **55:**277–296.

Levine, L., and Van Vunakis, H., 1970, Antigen activity of prostaglandins, *Biochem. Biophys. Res. Commun.* **41**(5)**:**1171–1177.

McGuire, J., McGill, R., Leeman, S., and Goodfriend, T. L., 1965, The experimental generation of antibodies to alpha-melanocyte stimulating hormone and adrenocorticotropic hormone, *J. Clin. Invest.* **44**(10)**:**1672–1678.

Nisonoff, A., 1984, *Introduction to Molecular Immunology,* Sinauer Associates, Sunderland, Massachusetts.

Ouchterlony, O., 1958, Diffusion in gel methods for immunological analysis, in: *Progress in Allergy,* Vol. V (P. Kallos, ed.), pp. 1–78, S. Karger, Basel.

Ouchterlony, O., and Nilsson, L. A., 1978, Immunodiffusion and immunoelectrophoresis, in: *Handbook of Experimental Immunology* (D. M. Weir, ed.), pp. 19.1–19.44, Blackwell Scientific Publishers, Oxford, England.

Reichlin, M., 1980, Use of glutaraldehyde as a coupling agent for proteins and peptides, *Methods Enzymol.* **70:**159–165.

Reichlin, M., Schnure, J. J., and Vance, V. K., 1968, Induction of antibodies to porcine ACTH in rabbits with non-steroidogenic polymers of BSA and ACTH, *Proc. Soc. Exp. Med.* **128**(2):347–350.

Reichlin, M., Nisinoff, A., and Margoliash, E., 1970, Immunological activity of cytochrome C. Enhancement of antibody detection and immune response initiation by cytochrome C polymers, *J. Biol. Chem.* **245**(5):947–954.

Steiner, A. L., Kipnis, D. M., Utiger, R., and Parker, C., 1969, Radioimmunoassay for the measurement of adenosine 3',5'-cyclic phosphate, *Proc. Natl. Acad. Sci. USA* **64**(1):367–373.

Stellwag, E. J., and Dahlberg, A. E., 1980, Electrophoretic transfer of DNA, RNA, and protein onto diazobenzyloxymethyl (DMB)—paper, *Nucleic Acids Res.* **8:**299–317.

Symington, J., 1984, Electrophoretic transfer of proteins from two-dimensional gels to sheets and their direction, in: *Two-Dimensional Gel Electrophoresis of Proteins: Methods and Applications* (J. E. Celis and R. Bravo, eds.), pp. 127–168, Academic Press, New York.

Towbin, H., and Gordon, J., 1984, Immunoblotting and dot immunobinding: Current status and outlook, *J. Immunol. Methods* **72:**313–340.

Tsang, V. C. W., Bers, G. E., and Hancock, K., 1985, Enzyme-linked immunoelectrotransfer blot (EITB), in: *Enzyme-Mediated Immunoassay* (T. T. Ngo and H. M. Lenhoff, eds.), pp. 389–414, Plenum Press, New York.

Vaitukaitis, J., Robbins, J. B., Nieschlag, E., and Ross, G. T., 1971, A method for producing specific antisera with small doses of immunogen, *J. Clin. Endocrinol.* **33:**988–991.

VanEldik, L. J., and Wolchok, S. R., 1984, Conditions for reproducible detection of calmodulin and S100 beta in immunoblots, *Biochem. Biophys. Res. Commun.* **124:**752–759.

Van Vunakis, H. V., and Langone, J. J., 1980, Immunochemical techniques, *Methods Enzymol.* **70:**1–507.

Williamson, A. R., 1978, Isoelectric focusing of immunoglobulins, in: *Handbook of Experimental Immunology* (D. M. Weir, ed.), pp. 9.1–9.31, Blackwell Scientific Publishers, Oxford, England.

Wood, D. M., and Dunbar, B. S., 1981, Direct detection of two crossreactive antigens between porcine and rabbit zonae pellucidae by radioimmunoassay and immunoelectrophoresis, *J. Exp. Zool.* **217:**423–433.

Zola, H., and Brooks, D., 1981, Techniques for the production and characterization of monoclonal hybridoma antibodies, in: *Monoclonal Hybridoma Antibodies: Techniques and Applications* (J. Hurrell, ed.), pp. 1–57, CRC Press, Boca Raton, Florida.

Chapter 9

Practical Methods for Laboratory Photography

I. INTRODUCTION

Although many investigators shy away from doing their own photography, there are many advantages for a student or investigator learning basic photographic methods in the laboratory. An excellent series of books by Ansel Adams (1980, 1981, 1983) are now available that give detailed descriptions of all aspects of basic photography. Among the most important reasons for learning photography for the laboratory are (1) cost effectiveness and efficiency, (2) immediate and complete data storage, and (3) accurate records of what is scientifically important.

A. Cost Effectiveness and Efficiency

Given the already escalating costs of research, one of the most significant expenses is recording data as well as preparing figures for publication. (This is particularly applicable for laboratories that need to photograph large numbers of polyacrylamide gels routinely.) Although professional services are sometimes essential, any student, technician, or investigator can readily learn practical and applied photography. With a minimal investment of photographic equipment, the savings to a laboratory are considerable (more than hundreds of dollars) over a short period of time. In addition to the costs incurred, the time required to send work out is generally inefficient and usually takes days, if not weeks. If basic photographic methods are used in the laboratory, it is possible to start from raw data and have publication ready prints in an afternoon.

B. Immediate and Complete Data Storage

A good motto to have in the laboratory is: Film is cheap, experiments and time are not. Even if an experiment is neither optimal nor of publication quality, it may turn out to contain valuable information that will be critical at some point. While some data (e.g., autoradiographs on x-ray film) can be stored for long periods of time, other information such as that obtained from Coomassie Blue- or silver-stained polyacrylamide gels may be lost as the gels age or break from handling or storage. If a laboratory sets up to take photographs routinely, the risk of losing important data is minimized.

C. Accurate Data Recording

In general, no one knows how to interpret scientific data better than the investigator carrying out the experiments. For this reason, the investigator will know what is most important and will be able to record this accurately when photographing data. A professional photographer will frequently sacrifice data for photographic aesthetics.

We have also found that the time spent in the dark room while printing and enlarging allows one to concentrate on the data in great detail. This practice has sometimes led to a reinterpretation of the original observations. Although the use of computerization of polyacrylamide gels (especially two-dimensional analyses) makes data recording of complex protein patterns more efficient, these methods do not replace the need for direct visual recording. Frequently, computer-generated data can be the result of an artifact in the gel pattern, a factor that can only be corrected by the visual interpretation of the original gel or of a high-quality photograph by the investigator.

II. CHOICE OF PHOTOGRAPHIC METHODS

A. Choice of Camera

The choice of camera will depend on the expense as well as the general photographic need.

1. 35-mm Camera

This is the most practical and convenient camera for small to medium-size laboratories. Since many people own a 35-mm camera, it is possible that they can use their own in the laboratory (with proper lens adaptation). The advantages of the 35-mm camera are that it is easy to load as well as to develop film in the laboratory using a dark bag (see Appendix 15). The major disadvantage is that the small size of the film itself (35 mm) may result in a grainy print if prints are enlarged too much. In general, however, closeup photographs of polyacrylamide gels are taken, and there are seldom reasons for excessive enlargement. This camera setup is the most cost effective and efficient for recording large numbers of experiments.

2. Other Cameras

A variety of other cameras, including those that have a 2 × 2-inch format cameras or 4 × 5-inch camera backs mounted on a Polaroid camera stand, will give slightly better image quality for photograph enlargement. We have found, however, that this is a time-consuming and expensive approach for photography of most gels in comparison with routine photography using a 35-mm camera. This is particularly true if large numbers of gel analyses are carried out. Photography of polyacrylamide gels using 4 × 5-inch film is described in detail by Tollaksen et al. (1984).

B. Choice of Lens

1. Choice of Lenses

Five basic types of lens have been described in detail by Adams (1980).

a. Normal Lens. The normal lens, such as the 55 mm, has been defined as one whose *focal length* is about equal to the diagonal of the film format. This lens will have an angle of view of 50–55%, which is similar to that of normal human vision. (The *focal length* of a lens refers to the distance from the rear nodal point of the lens to the plane where subjects at infinity come into focus.)

b. Short Focal Length Lens. This type of lens (>65% angle of view) projects a wide area of subject on the film and is therefore referred to as a wide-angle lens. It has a greater depth of field and is commonly used for photography of broad landscapes.

c. Long Focal Length Lens. This is used when photographing distant objects or scenes because images are enlarged on the film.

d. Zoom Lens. The zoom lens has been designed so that the effective focal length can be varied through a specific range, thus altering the image size. A good zoom lens can therefore replace a variety of focus length lenses with a single lens.

e. Macro Lens. This is the lens of choice for photography of most polyacrylamide gels. The term *macro* refers to a lens that can focus at close subject distances. Because some macro lenses (about 100 mm focal length for 35-mm cameras) are intended for use with a supplementary bellows, they may not focus to infinity. Furthermore, because a relatively large magnification of the image also magnifies the movement of the camera during exposure, small movements will affect image resolution. For closeup photographs of polyacrylamide gels, we have used a Minolta MD-Macro Rokkor-X (1 : 35, f = 50 mm) lens with consistently good results.

2. Controlling Light by Lens Aperture

The aperture of a lens works similar to the opening and closing of the pupil of an eye. The size of the opening of the aperture is measured on a standard scale of numbers called f stops. (*Note:* The letter f has nothing directly to do with focusing or focal length in this instance.) The numbers of f stops may range from f/1.0, which admits the largest amount of light, to f/64, which admits the smallest amount of light. For example, an f stop of 64 admits less than 1/2000th of the light (see the discussion in *Light and Film,* Time–Life Books).

C. Choice of Film

1. Introduction

Various types of films have emulsion developed to respond differently to light. A summary of types of film is given in Table 9.1. Every film has a designated rating, the ASA, which is the exposure index designated by the American Standard Association. The ASA defines numerically the

Table 9.1. Film (35-mm) Commonly Used in Laboratories

Film	Use
Kodak Panatomic X	Polyacrylamide gels, autoradiography
Kodak Tri-X	Polyacrylamide gels, autoradiography
Kodalith	Black and white print or line drawings
Direct positive movie film (Eastman MP5360)	Direct-positive slides
Kodachrome 64 or 200 (daylight)	Color slides of polyacrylamide gels

speed of the film according to the amount of light needed to produce a normal image. The higher the number of the ASA (designated a *fast film*), the less light will be needed for illumination. The lower-rated ASA films (*slow films*), however, will give sharper, less grainy images. In addition to speed and graininess of films, there are also variations in color sensitivity (see the discussion in *Light and Film,* Time–Life Books).

2. Formation of Image on Film

The details of image formation on film is also given in *Light and Film* (Time–Life Books). In summary, black and white film is made up of a number of layers. The surface layer consists of a protective coating to prevent scratches on the emulsion. The adjacent emulsion layer consists of gelatin and silver bromide crystals in which the image will be formed. The emulsion is bound to a firm base support (usually cellulose-acetate). The final layer consists of an antihalation layer, which will prevent the light from reflecting back through the emulsion to cause halos around the film images.

When a photon of light hits silver bromide crystals in the light emulsion layer, the formation of the image will begin. This photon will transfer energy to the electrons of the bromide ions, raising it to a higher energy state. This negatively charged electron will then interact with a sensitivity speck (the impurity in the crystal, such as silver sulfide) as well as the positively charged free silver ion. As additional photons of light strike the bromide ions, more silver ions migrate to the sensitivity speck and are transformed to atoms of silver metal. The presence of several metallic silver atoms at a sensitivity speck will give rise to the *latent image,* which itself is too small to be visualized but which forms a visible image when developing chemicals are used to convert the entire crystal.

Because different films have different sizes of silver bromide crystals, this will affect the graininess of the film. Films containing large silver bromide crystals will produce coarser-grain film negatives, but these have high sensitivity to light (fast films) because a large crystal does not need any more light to form a latent image than a small crystal, but it yields more metallic silver when developed. In summary, one should use a slow film (e.g., Panatomic X) if maximum sharpness of image and minimum graininess is desired (from discussions in *Light and Film,* Time–Life Books).

3. Choice of Black and White Film for Photography of Polyacrylamide Gels

Because it is desirable to obtain the highest resolution photographs of your polyacrylamide gels, it is generally practical to use slower-speed film, such as Panatomic X film. Tri-X film may also be used with success for some purposes. Frequently, a high-contrast negative is needed for line drawing or data summary. For this purpose, a high-contrast film such as Kodalith film is best (Kodalith negatives can also be used to prepare colored slides, i.e., white print on blue background, which are popular for scientific presentations). Direct-positive slides can also be made of polyacryalmide gels or line drawings if a special film is used.

4. Color Film for Photography of Polyacrylamide Gels

In black and white photography, the major concern is with light intensity and contrast. In color photography, however, the color quality of the light source is critical. For this reason, there are two types of emulsion for color film. The term *color temperature* has been used to describe the color of light, which may vary from light source to light source. The color temperature of daylight (blue-violet end of color spectrum) is very different from that of tungsten light (yellow-orange spectrum). The emulsion on daylight film is balanced for bright sunlight and can only be used with tungsten light if a blue filter is used, which reduces the speed of the film (Sussman, 1973).

Because the light source of lamps used in light boxes may vary, it is important to determine which film will give optimal results. We have found that daylight film is best with most high-intensity light boxes used for photography of gels. We routinely use Ektachrome 100 film for taking color slides of gels. These can also be used to print directly onto Cibachrome positive paper (see Section VII).

F. Choice of Filters

In general, there are at least two reasons why filters are used in photography: (1) to increase the contrast between tones; and (2) in color photography, to enable the emulsion to register various colors more closely to those than the eye sees.

The types of filters that are available include plain gelatin film, as well as plastic or glass filters. Although glass filters are generally more uniform and more accurately ground, the film or plastic filters are usually adequate for routine photography of polyacrylamide gels.

1. Filters for Use with Black and White Photography

Yellow, orange, or red filters can be used to photograph Coomassie blue-stained gels. The yellow filter will absorb ultraviolet light and some of the blue violet light so that the blue will be darkened. The orange filter will absorb ultraviolet, violet, and most of blue light, whereas a red filter will absorb these colors as well as green light. These can therefore be used to darken the image of blue-stained proteins in gels.

2. Color Silver-Stained Gels

It is common for investigators to avoid the use of color-silver staining methods because they cannot afford the publication costs of color plates. We have found, however, that excellent resolution can be obtained using black and white photographs of color-stained gels. Since the polychromatic stained gels contain all shades of reds, blues, oranges, yellows, and greens, it is difficult to use filters to optimize all color intensities. In general, we have had the best results with Panatomic X film without filters. With the proper exposure, the resolution of color spots can easily be optimized.

III. BULK LOADING FILM

If photography of polyacrylamide gels is a routine laboratory practice, it is advisable to use bulk rolls of film. The savings from the use of bulk film is substantial. The equipment and methodology needed for this simple procedure are outlined in Appendix 16. A further advantage of this method is that the exact amount of film can be selected; thus film will not be wasted as it frequently is if commercial rolls of film are used.

IV. PHOTOGRAPHING POLYACRYLAMIDE GELS

It is important to note when loading a bulk loader with a new roll of film that most darkroom safe lights are *not* safe for 35-mm film. Caution must therefore be taken not to expose the bulk roll of film. A dark bag such as that described in Appendix 15 is easy to use for loading bulk film.

A. Photographic Setup

In order to photograph polyacrylamide gels, it is necessary to use a light box set up with a camera stand (see Appendix 15). A 35-mm camera with a macro lens (as described in Section II.B) can be used to photograph most gels.

ethylene)-coated papers have properties that prevent chemicals from soaking into the paper fibers, hence speed processing (Adams, 1983).

2. Paper Weights and Grades

Papers can come in a variety of weights determined by the thickness and bulk of the paper base. Thicker papers (such as double weight) are usually preferable because they better withstand handling and processing and lie flatter after drying on print racks (Adams, 1983).

Paper grades are numbers that indicate the contrast of the paper. The softest papers are numbered grade 1 or 0, as compared with the hardest papers, which range to 5 or 6 depending on the manufacturer. In general, the type of paper used should match the density range of the negative. For example, a high-contrast negative (such as a Kodalith negative) will require a paper of long exposure scale (grade 0–1), while a low-contrast negative (such as a Panatomic X negative of polyacrylamide gel) will require a shorter-scale paper (grade 4–6). Alternative papers are those that contain two different emulsions, one of low and one of high contrast. These are known as variable contrast papers (e.g., Kodak Polycontrast or Ilford Multigrade). Because these papers will have a different exposure

Table 9.3. **Process for Developing Individual Black and White Prints**[a]

Process	Reagent	Time	Comments
Develop	Dektol	2–3 min	Immerse print quickly or smoothly face down and agitate.
Stop bath	Dilute acetic acid or Kodak stop bath	30 sec	Immerse, agitate, and drain.
Fixative	Kodak rapid fixer	5 min	Be sure that prints do not stick together.
Hypo clearing	Kodak hypoclearing agent	3 min	This step helps eliminate time for water washes.
Wash	Running water	30 min to 1 hr	Prints may be left overnight in distilled water if they cannot be dried that day.
Drying	—	—	Air-dry resin-coated papers; use print dryer for non-resin-coated papers.

[a]Modified from Adams (1983).

scale depending on the color of the enlarging light, filters can be used to control the image contrast. Thus, a single paper can be used to replace the use of different grades of papers.

B. Print Exposure

The time of exposure of the negative image will depend on the negative, paper, light source, and light aperture of the lens. The optimal time exposure can be determined by making a test strip in which different time exposures or f stops are used on a single photograph. If there are light or dark regions on the negative that you wish to even out on the paper, it is possible to haze or dodge the print by moving your hand or other object rapidly back and forth between the light source and the paper. It is essential to take care that the scientific data are not altered by this procedure.

C. Print Development

The basic process of developing prints is similar to that of film development. The process for developing individual prints in pans is summarized in Table 9.3.

VI. PRINTING PHOTOGRAPHS FROM NEGATIVES

Although printing one's own photographs is cost effective, it is frequently necessary for investigators to have prints made professionally. If the investigator has taken care to prepare a high-quality negative that has the proper contrast and detail, it should be easy to have prints made professionally as well as to print them oneself.

A. Equipment

A well-equipped darkroom is necessary for printing high-quality photographs from negatives. This equipment includes a photographic enlarger with approved lenses and negative carriers, paper easels, focusing magnifiers, timers, and trays for print developing. Detailed descriptions of this type of equipment are outlined elsewhere (see Adams, 1983).

VII. DIRECT CONTACT PRINTS OF COLOR-BASED SILVER-STAINED POLYACRYLAMIDE GELS

It is not feasible to publish color photographs of all gels that have been stained with the color-based silver stain; however, it is generally preferable to have color plates on poster presentations at scientific meetings. For this reason, we have worked out a simple, inexpensive method for preparing color contact prints of the gels themselves (Dunbar *et al.*, 1983). (This method is outlined in detail in Appendix 16.) It simply involves the placement of the gel on a glass plate, which is then placed onto a sheet of color paper (Cibachrome) and is then exposed with a print enlarger (with appropriate color filters). The paper is then developed. A high-quality color print of the gel is ready within just a few minutes.

REFERENCES

Adams, A., 1980, *The Camera,* New York Graphic Society–Little, Brown, Boston.

Adams, A., 1981, *The Negative,* New York Graphic Society–Little, Brown, Boston.

Adams, A., 1983, *The Print,* New York Graphic Society–Little, Brown, Boston.

Dunbar, B., Bundman, D., and Dunbar, B. S., 1983, A rapid laboratory method for developing color contact prints of silver stained polyacrylamide gels, *Electrophoresis* **4:**258–259.

Sussman, A., 1973, *The Amateur Photographer's Handbook,* 8th rev. ed., Thomas Y. Crowell, New York.

Time-Life Books, 1970, *Light and Film,* Time–Life Books, New York.

Time-Life Books, 1970, *The Camera,* Time–Life Books, New York.

Tollaksen, S. L., Anderson, N. L., and Anderson, N. G., 1984, Operation of the ISO-DALT System, 7th ed., Argonne National Laboratory, #ANL-BIM 84-1, Argonne, Illinois.

Chapter 10

Troubleshooting and Artifacts in Two-Dimensional Polyacrylamide Gel Electrophoresis

I. INTRODUCTION

Although polyacrylamide gel electrophoresis (PAGE) methods have been used extensively over the past two decades and the use of two-dimensional electrophoresis systems are becoming increasingly popular, the data generated by these techniques are frequently far from optimal. Not only are numerous problems associated with this technology that are frustrating and discouraging to students and technicians as well as to investigators just starting this methodology, but the data that are eventually published are frequently not interpretable or reproducible in other laboratories. Most of these technical difficulties can now be overcome if the nature of the problems encountered is understood. Some of these problems were previously addressed (VanBlerkom, 1978; Bravo, 1984). The problems discussed in this text are those that students and technicians have commonly encountered in my laboratory over the past few years.

It is further necessary to appreciate that reproducibility in your gel system does not guarantee that your technique is optimized. It is recommended that an investigator new to this technology begin these techniques with a sample such as human plasma or serum. Because numerous laboratories have characterized these protein patterns, it will be possible for investigators to determine immediately whether their techniques are optimized.

II. REAGENT QUALITY AND EQUIPMENT MAINTENANCE

A. Water

Although every laboratory is generally equipped with a still for generating distilled water or with other deionizing water systems, we have found that optimal water quality must always be maintained for optimal results. Because contaminating ions and organics can interfere with isoelectric focusing (IEF) and gel polymerization as well as with staining reactions, it is critical that an optimal source of water be used for all methods.

Two general considerations should be taken into account when choosing an acceptable water system for laboratory use with electrophoresis systems. First, the quality of the source of water will vary considerably from region to region. More stringent water purification methods may be needed in areas in which water quality is poor. Second, the quantity of water to be used must be determined, as some water purification systems (e.g., triple-distilled water units) will not process water rapidly enough to keep up with daily needs. Since many of the multiple gel electrophoresis systems require large volumes of buffers, this can be a significant problem.

We have found that a multiple-filter deionizing system such as that supplied by Millipore (mixed-bed $+,-$ resin made by Rohm Haas) is excellent for most methods described in this text. This system is also capable of providing large volumes of water rapidly without long-term storage in containers that might permit bacteria to grow or that might leech organic compounds, e.g., many plastic bottles.

B. Chemicals

1. Ampholyte

One of the major variables in two-dimensional PAGE is the composition of ampholytes. The methods used to synthesize ampholytes vary and the complexity of the synthesis procedures does not guarantee a reproducible product, leading to a great deal of variation in the source and batch of ampholytes. Different brands or lots of ampholytes may interact with detergents such as SDS or with contaminating substances frequently found in detergents. It may therefore be necessary to test ampholytes for use with the solubilization buffer since such interactions may cause IEF gel breakage or streaking in the second dimension. It is frequently necessary to mix different batches of ampholytes having differing ranges of pH. It is therefore advisable that investigators determine which is the best reproducible source of ampholytes for their needs. It is also essential that some form of standardization procedure be used when first establishing this method in your laboratory or when using a new source or mixture of ampholytes. This approach will enable you to determine whether your isoelectric focusing technique is optimal and to compare your data with those of other laboratories. (The internal standards are far superior to the measurement of pH in gels in assessing the reproducibility of the pH gradient generated by a particular lot of ampholytes) (see Fig. 3.3, Chapter 3; see also discussions by Anderson and Hickman, 1979; Tollaksen *et al.*, 1981).

2. Acrylamide

Because acrylamide can be an expensive item if large numbers of polyacrylamide gels are used, it is generally cost effective to try to obtain an inexpensive source of acrylamide. It is critical, however, that this consideration never be at the expense of your experiments. While a more inexpensive (i.e., less purified) form of acrylamide may be filtered as described in Appendix 4 and can sometimes be acceptable for second-dimension slab gel electrophoresis, it is not advisable to use this for IEF. It is advised that a high-quality electrophoresis-grade acrylamide be used if you are just initiating these methods or if you are only running small numbers of acrylamide gels, so that expense is not an issue. Once you have all the techniques mastered, it may then be practical to set up a filtering system that will permit you to switch to more inexpensive brands of acrylamide.

3. Buffers

In general, we have found that most of the commercial sources of reagents such as Tris and glycine, which are used to make up electrolyte solutions for electrophoresis, are consistent. However, to ensure high-quality results when you are just starting the techniques, we recommend obtaining these reagents from sources that specialize in electrophoresis quality reagents.

It is also critical to take into account several factors when determining the pH of buffers. Because some pH meter electrodes do not work with Tris buffers, significant errors can be made if incorrect electrodes or noncalibrated pH meters are used. Because pH variation may result from variation in salt concentration, the improper adjustment of buffers can alter the current during electrophoresis (Lane, 1982). If you encounter problems during electrophoresis, it will be necessary to evaluate critically the methods you are using to adjust your pH values.

4. Detergents

The quality of detergents for sample solubilization as well as for use in polyacrylamide gels is very important. For example, contaminants in sodium dodecyl sulfate (SDS) may adversely effect protein and will interfere with IEF solubilization and can inhibit polyacrylamide gel polymerization. If SDS does not go into solution readily, it is not advised that this source be used, as it may interfere with gel polymerization. If electrophoresis-quality detergents are obtained, this problem can usually be avoided. While there are currently limited sources of nonionic detergents such as Nonidet P-40 and CHAPS, the quality is dependent on the supplier.

C. Equipment Maintenance

For optimal results, it is always necessary to keep equipment dust free. If you are using multiple electrophoresis gel systems with recirculating buffer units, it is also important to be aware that bacteria, algae, and other organisms can grow in tubing lines. We have found that tanks that are not in daily use should be wiped down with alcohol solutions and washed thoroughly with SDS and with several changes of water before carrying out electrophoresis. These procedures are especially important if silver-staining techniques are being used routinely, because organics or other contaminants that leech out of Plexiglas, rubber, glues used to repair tanks, and so forth can give silver-staining artifacts as well as high background staining (see Fig. 10.1).

Power supplies should be monitored routinely. We recommend using power sources that have both amperage as well as voltage monitors, so that you can be sure that your gel systems are running properly. For example, when you first turn on the power for your IEF gels, the amperage will read 5–15 mA, depending on the number of gels in the electrophoresis chamber, but at "equilibrium" following IEF it will read 0. For second-dimension SDS electrophoresis, the amperage reading will be high initially if constant voltage is used but will then drop during electrophoresis. Failure to demonstrate these changes in amperage indicates the need to adjust your electrophoresis apparatus.

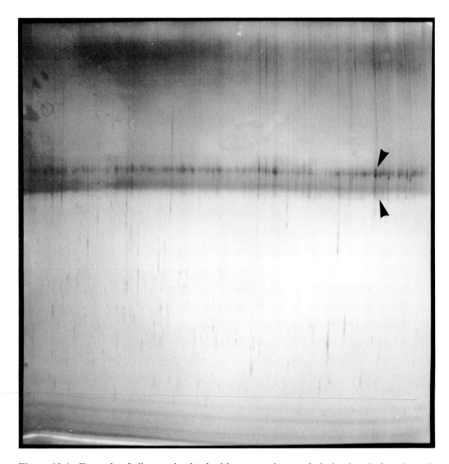

Figure 10.1. Example of silver-stained gel with no protein sample in isoelectric focusing gel. The upper dark staining region of the gel was due to organics in the result of improper washing of the electrophoresis tank prior to electrophoresis. This gel also contains the common silver artifacts (arrows) that always occur at the 69,000- and 55,000-M_T range. These are associated with the presence of β-mercaptoethanol in the IEF gel equilibration buffer. See discussion in text.

III. ISOELECTRIC FOCUSING

A. Sample Solubilization

Sample solubilization is perhaps the most critical parameter for obtaining reproducible and high-resolution protein patterns in two-dimensional PAGE. A variety of solubilization methods may have to be tried in order to arrive at the best system for your proteins. The following problems are those that may commonly arise if improper solubilization or sample preparation methods are used.

1. Salt and Nonamphoteric Ion Effects

One of the most frequently encountered problems in IEF is the presence of nonamphoteric ions that may alter the apparent pI of proteins. Since many samples will have to be concentrated prior to analysis by electrophoresis, one must use methods that will concentrate the protein but that will not increase the salt concentration. This can be achieved by the use of centrifuge concentrators or microdialysis methods. If long-term dialysis is used on some complex protein mixtures, however, there may be extensive proteolysis, so one must determine whether this is acceptable for your sample. Concentration of proteins by precipitation, e.g., ammonium sulfate fractionation, ethanol, or trichloroacetic acid (TCA) precipitation, may also give rise to problems, and the latter two methods may alter charge or solubility properties of proteins. Figure 10.2 illustrates a serum protein pattern that was obtained when a high concentration of ammonium sulfate was present in the sample. The apparent pI range of these proteins is dramatically altered in the presence of this salt.

2. Ratio of Detergent to Protein Used for Protein Solubilization

Guidelines for solubilizing proteins for IEF or for SDS–PAGE commonly used by most investigators are presented in Appendix 2. If too little detergent is used, inadequate solubilization may occur. Since it has been estimated that up to 3 mg SDS will bind 1 mg protein, it is recommended that you be sure that you have at least this concentration of detergent in your sample solubilization buffer.

3. Nucleic Acids

Because nucleic acids are also charged molecules that can interact with proteins as well as alter apparent pI ranges, they can cause problems. Furthermore, nucleic acids also stain with silver-staining procedures, making protein interpretation more difficult. Some of these problems can be eliminated or minimized if adequate sample-preparation methods are used. Investigators who are using radiolabeled methods to identify proteins frequently use nucleases to break down nucleic acids (Garrels, 1979). While these enzymes are adequate for autoradiography, they have the following disadvantages: (1) the commercially available nuclease prepa-

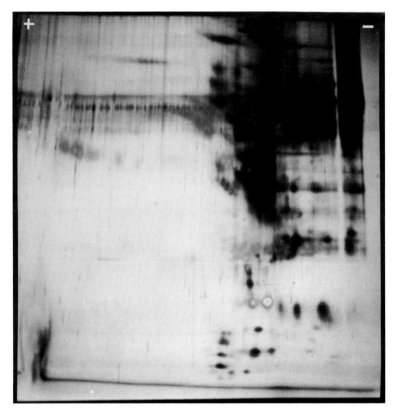

Figure 10.2. Protein pattern of serum sample containing high concentration of ammonium persulfate prior to solubilization and isoelectric focusing. The apparent pH gradient is altered, and much of the protein does not enter the basic end of gel.

rations are seldom pure and may contain dozens of proteins if analyzed by two-dimensional PAGE and silver staining methods and could therefore complicate protein patterns even more; (2) nuclease preparations may contain proteases, which will alter protein patterns; and (3) nucleases may digest nucleic acids to small enough fragments that they cannot be removed by ultracentrifugation, as are nondigested nucleic acids. If you have a sample such as a nuclear matrix preparation that can be washed to remove excess nucleases as well as nucleic acid fragments, this method can be used (Simmen *et al.*, 1984). If these nucleic acid fragments cannot be removed from the sample, however, their wide range of charge and molecular-weight range will result in very high silver-staining backgrounds, as illustrated in Figure 10.3.

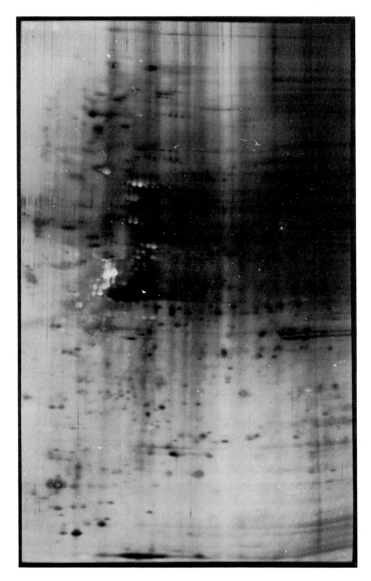

Figure 10.3. Example of intense background staining due to nucleic acids. Even with this contamination, however, it is possible to visualize many distinct proteins if the color-based silver stain Gelcode is used.

The most effective method for the removal of nucleic acids from samples to be used for silver staining is simply to ultracentrifuge the sample (100,000–200,000 $\times$ g) for 1–2 hr after solubilizing the sample. Since the nucleic acids will be denatured but not fragmented, most them can easily be removed by this procedure. The possibility exists, however, that some proteins may interact with nucleic acids even in the presence of SDS and urea. If your proteins of interest are in this category, this will not be the optimal procedure for your purposes.

B. Problems with Gel Polymerization

Failure of gels to polymerize when you are casting them in tubes has several possible causes.

1. Too little of the catalysts was used [ammonium persulfate or N′N′N′-tetramethylethylenediamine (TEMED)]: Check concentrations and increase concentrations of each as required to obtain polymerization—usually can increase by 10% increments for adequate results. This can also be caused by hydration of your ammonium persulfate. It is therefore recommended that you use fresh, electrophoresis-quality ammonium persulfate. If you make up a large batch of stock ammonium persulfate and freeze aliquots and store in dark bottles as recommended in Appendix 3, this should rarely be a problem. Finally, if you add water to the ammonium persulfate and listen to it carefully, it should "snap, crackle, and pop" as it is going into solution. If this does not happen, the chances are that you do not have an adequate quality of reagent.

2. Acrylamide or bis-acrylamide concentrations may be inaccurate: Check calculations of stock solutions. It is not advisable to increase concentrations of either of these reagents. If stored improperly, hydration of acrylamide may result in improper weight measurements for stock solutions. This should be taken into account before making up solutions. Finally, if inferior quality of either acrylamide is used, contaminants may interfere with gel polymerization.

3. Glass tubes were not washed adequately and have detergent or other residues that interfere with polymerization: Wash tubes well in chromic acid and rinse well with a high-quality water. For best results, tubes should be air or vacuum dried.

4. Acrylamide solution was not adequately degassed: Since O_2 can interfere with polymerization, it is important to degas solution

adequately to ensure uniform gel polymerization as well as to remove other nonamphoteric ions.

C. Problems with Removal of Gels from Tubes

Occasionally, IEF gels are not readily removed from tubes; by contrast, they may fall out of tubes either before or during IEF, which is usually attributable to the following:

1. Tubes were inadequately washed: Rewash the tubes in chromerge, distilled, or deionized water and vacuum-dry the tubes.
2. Improper-sized glass tubes have been used: If the size of tubes described in Appendix 3 is used for this method, this problem should never occur. Slightly larger tubes can sometimes be used for isoelectric focusing, but these gels are generally more difficult to remove from the tubes. It is sometimes possible to siliconize tubes, but this is not generally advisable, since gels may fall out of the tubes during IEF and could result in considerable equipment damage.
3. Too little Nonidet P-40 was added to the acrylamide solution, making gels difficult to remove: It is advisable to cut off the end of the Eppendorf pipette tip so that this viscous detergent can be accurately measured. Be sure to wipe off the excess detergent from the outside of the pipette tip so that excessive detergent is not added. On several occasions, the use of excessive detergent in the IEF gel has resulted in gels falling out of the tubes.

D. Gel Breakage during Isoelectric Focusing

This has been a very common problem and one of the most frustrating encountered with this technology. We have found several factors to contribute to this problem.

1. Samples contain high salt concentrations such as lyophilized buffer salts and ammonium sulfate: These will not only interfere with IEF but will also result in gel breakage. All samples should be prepared using low salt concentrations, as described in Appendix 2 on sample preparation.
2. Different lots of ampholytes will frequently have a "hole" in the pH gradient, where no ampholytes will migrate during IEF: This will result in local heating and damage to the gel matrix in that

area. This problem can usually be overcome by mixing ampholytes from two sources. Because different methods are used for ampholyte preparation by different companies, the chances of filling in the "hole" in the gradient are good. The disadvantage to this approach is that you are adding yet another variable (e.g., two different lots of ampholytes) to the system. It is therefore advisable to mix a larger lot of ampholytes for consistency and to use internal charge standards initially to determine the relative pH range of your ampholyte mixture.

3. Your protein concentration is too high for the size of the gel: Gel breakage is particularly a problem if impurified or abundant proteins are used and they concentrate in one part of the gel (i.e., at their isoelectric point). Because silver stains require much less protein, this problem can frequently be overcome.

E. Irreproducible Isoelectric Focusing Patterns

A great number of factors can cause irreproducibility in IEF patterns. Many of these are discussed elsewhere in this text. The following list summarize the major sources that contribute to gel irreproducibility.

1. Source and batch ampholytes: Since most investigators have to depend on commercial suppliers for reliable quality and consistency of ampholytes, there is little one can do to control this factor. It is sometimes possible to obtain a select lot of ampholytes from a given supplier if large quantities of ampholytes are to be used, but this is not practical for most users, since some sources of ampholytes cannot be stored for long periods of time without adverse effects. An example of inadequate focusing due to an inferior lot of ampholytes is shown in Figure 10.4.

2. Contaminating ions in sample or in reagents: Because nonamphoteric ions such as salts or carbon dioxide can alter the establishment of the pH gradient during IEF or alter the solubilization conditions, these factors can play a major role in reproducibility of IEF patterns.

3. Volt-hours used for isoelectric focusing: This factor can be a major one for many reasons. Since all proteins will not reach their relative pI range in the same length of time, it is important to optimize and standardize conditions best suited for adequate and reproducible isoelectric focusing. The time and voltage needed for isoelectric focusing will depend on the length of the gel as well as

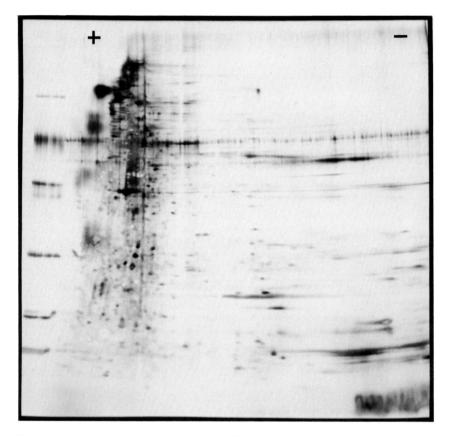

Figure 10.4. Typical protein pattern of total cellular protein resulting from an inadequate ampholyte batch or range, causing most of the proteins to pile up at acidic end of gel. Cathodic drift is also apparent in this gel pattern.

conditions used for solubilization. We have also observed that it is not adequate simply to use a consistent number of total of volt-hours as the endpoint. For example, carrying out isoelectric focusing for 15 hr at 800 V may result in different patterns than 500 V for 24 hr. It is therefore necessary to optimize and standardize the volt-hours used for isoelectric focusing. In general, we have found that the use of higher voltage for shorter periods of time results in sharper resolution of proteins in the isoelectric focusing gel. If a nonequilibration (NEPHGE) gel system is used, however, multiple time points may have to be chosen to optimize for separation of proteins, as shown in Figure 10.5.

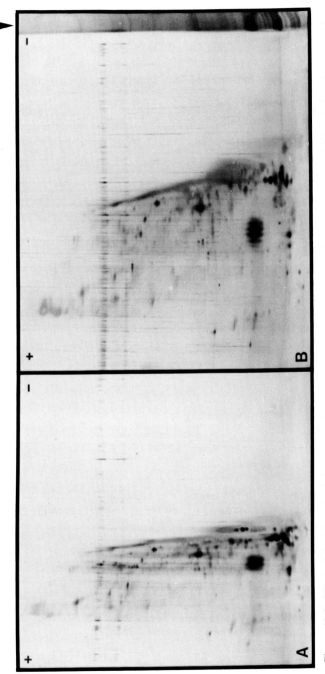

Figure 10.5. Effect of duration of volt-hours used for electrophoresis of proteins with the nonequilibration isoelectric focusing system for resolution of basic proteins. Sample of macronucleus of *Tetrahymena thermophila* provided by Dr. Dave Ellis, Department of Biochemistry, Baylor College of Medicine. (A) Isoelectric focus (first-dimension) carried out for 1400 volt-hr. (B) Isoelectric focusing carried out for 2100 volt-hr. Second dimension (arrow) is same sample run in one dimension.

4. Cathodic drift: Cathodic drift can cause major distortions during isoelectric focusing. This will result in loss of resolution of basic proteins (see Fig. 10.4). This happens because the cathodic end components of the gel buffer are continuously passing into the reservoir buffer. The quality of ampholytes, as well as the number of volt-hours and other conditions used for focusing, will affect the degree of cathodic drift. Cathodic drift may be reduced by optimizing focusing conditions, e.g., increasing the ampholyte concentrations in the gels, or by adding these to the electrode

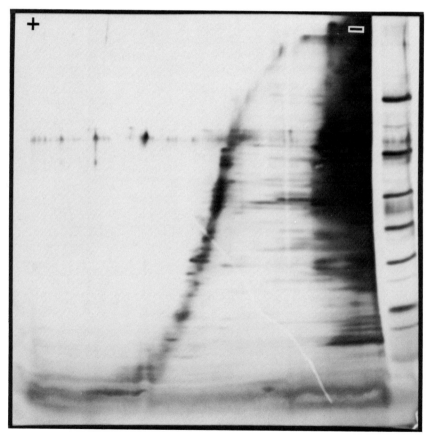

Figure 10.6. Typical protein pattern obtained when air bubble is present in tube above sample during electrofocusing.

buffer. (The latter is an expensive solution to this problem.) Alternatively, amino acids can be added to the electrode buffers.

5. Amount of protein used for isoelectric focusing: Since proteins have an inherent pI, they can frequently act as ampholytes themselves and can alter the distribution of the ampholytes. This is particularly pronounced if excessive amounts of proteins are used or if an abundant protein is present in the sample. It is always critical that these factors be taken into account when interpreting gel patterns. Again, this is where the use of internal charge standards is invaluable because alterations in the pH gradients caused by different protein samples can easily be evaluated.

6. Chemical modification of proteins prior to IEF: Many chemical labeling procedures or protein-purification procedures can alter charges of proteins. For example, heating in urea can cause carbamylation of a protein, which may dramatically alter its charge properties. It is therefore necessary to take into account all chemical procedures used to prepare samples in order to determine whether proteins will be altered, migrating differently at different times.

7. Inadequate solubilization: If too little solubilization buffer is added to the protein sample or if insufficient time of solubilization is used, many of the proteins may not even enter the IEF gel. It is common to see proteins aggregated at the basic end of the second-dimension gel.

8. Air bubbles in polyacrylamide or sample loading: If air bubbles are formed in the IEF polyacrylamide gel or if they are formed while the sample is being loaded, they can dramatically interfere with focusing (see Fig. 10.6). If tube gels are cast by capillary action, as shown in Appendix 3, this should not be a problem. It is usually possible to remove air bubbles after sample loading by placing a long needle down the side of the tube to release the air.

IV. SODIUM DODECYL SULFATE SLAB GEL ELECTROPHORESIS

In general, there are fewer problems with the mechanics or the reproducibility of the second-dimension SDS slab gel electrophoresis as compared with the first-dimension IEF procedures. This is particularly true if multiple gel-casting equipment designed for casting gradient gels is used. Nevertheless, the following problems may arise.

A. Gel Polymerization

Failure of polyacrylamide to not polymerize within 10–30 min after the gradient is poured could be attributable to any of several factors. These problems are the same as those described above for isoelectric focusing gel polymerization (Section III.B) and should be corrected in the same manner.

1. Catalyst (TEMED and Persulfate) Concentration

For the sake of reproducibility, gels should polymerize at a constant rate, as discussed by Chrambach and Rodbard (1981). These investigators have found that analytical-scale polyacrylamide gels (0.6-cm diameter) polymerized within 10 ± 2 min exhibit a maximum degree of conversion of monomers to polymer and a maximum chain length. Because the degree of conversion of monomers to polymer and the maximum average chain length are antagonistic, an increase in catalyst concentration will increase the percentage conversion but will decrease the average chain length. In practice, therefore, the lowest possible catalyst concentration causing polymerization within 10 min is recommended. If polymerizations are either too slow or too rapid, the catalyst concentrations will have to be changed gradually until optimal conditions are obtained. (*Note:* While Chrambach and Rodbard recommend the combination of photopolymerization with riboflavin in the presence of persulfate and TEMED, we have found that this is neither practical nor necessary if the multiple-gradient gel-casting systems are to be used to prepare gels.)

Finally, it should be noted that the source batch and age of catalyst reagents may alter the polymerization rate. These should be standardized where possible.

2. Acrylamide or Bis-Acrylamide Concentrations

The concentrations of both acrylamide and bis-acrylamide are critical for polymerization as well as for the formation of desired pore sizes (see Chapter 1, Section VIII). Care should be taken to use high-quality acrylamide (as recommended in Appendixes) and to store reagents adequately in order to avoid hydration of reagents.

3. Glassware and Plastic

Many substances are capable of inhibiting polymerization. The precast gel plates with permanent glass spacers (are available from Health

Products, Inc., or from Electronucleonics, Inc.) are optimal for casting acrylamide gels because they can be easily cleaned. Care should be taken to wash plates thoroughly each time (e.g., soak in SDS solution, rinse well with distilled H_2O and then 70% ethanol) in order to remove grease or contaminants that might also bring about irregularities in polymerization.

4. Inadequate Degassing

This problem can cause irregularities of polymerization or alter polymerization times. If large volumes of acrylamide are to be used such as for multiple gel-casting chambers, this becomes even more critical.

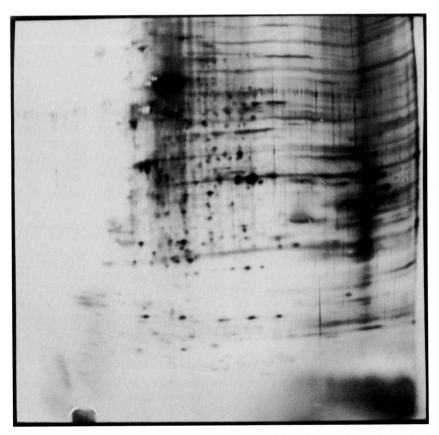

Figure 10.7. Typical pattern is formed due to improper mixing of acrylamide during gradient formation.

Thorough degassing with a high-vacuum pump or lyophilizer is recommended for this procedure.

5. Irregular Gradient Formation

The irregular formation of acrylamide gradients as shown in Figure 10.7 is a common problem associated with this technology. This can be caused by improper mixing of acrylamide due to inadequate gradient-casting equipment or to improper use of equipment. Mixing speeds for combining acrylamide solution should be both optimized and standardized for all equipment used in order to obtain reproducible gradients. Speeds should also be optimized to ensure good mixing without production of excessive numbers of air bubbles that may affect the acrylamide gradient after it is formed in the gel-casting chamber.

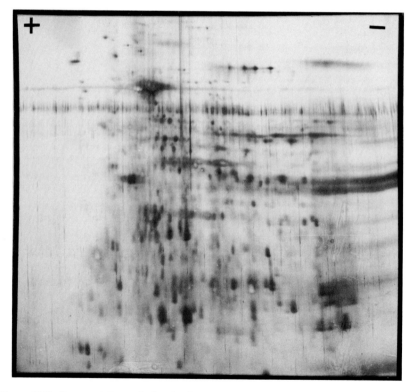

Figure 10.8. Poor protein resolution in the second dimension when plates are too tightly compressed against acrylamide during electrophoresis.

The time it takes to pour a gradient gel chamber is also important. If casting the gels takes too long, the acrylamide may start to polymerize before the gradient is established. We have found that 5–10 min is generally an optimal time in which to fill a gradient chamber. If glass gel plates are soiled, the contaminants may interfere with even polymerization. This can cause irregular protein patterns due to distortion of polyacrylamide matrix. A common pattern is one we refer to as "beetle bark" because of its resemblance of tree bark infected with beetles.

6. Mechanical Problems during Electrophoresis

Irregularities in electrophoresis may occur if gels are allowed to heat unevenly to a high amperage. If used so that the electrophoresis tank buffer does not maintain an even temperature, some proteins may migrate more quickly than others. This is a real problem if electrophoresis of gels is carried out in chambers in which gels are placed on their sides.

Another problem with protein resolution will be encountered if glass plates are compressed too tightly during electrophoresis. This can cause poor resolution of proteins, as illustrated in Figure 10.8.

V. PROBLEMS WITH STAINING GELS

A. Coomassie Blue Staining

1. Sensitivity of Protein Stains

There are different types of dyes, which are all referred to as Coomassie Blue (see discussion in Chapter 4). The source of dye as well as the lot of the dye may therefore alter the staining properties. In general, electrophoresis-grade qualities of Coomassie Blue such as those commercially available from BioRad Chemical Co. are adequate. While some investigators reuse stock solutions to stain multiple gels, it is advisable to use fresh stain if optimum sensitivity is required.

2. Smears on Top of Gels

This stain should be allowed to go completely into solution followed by filtration with Whatman #2 filter paper followed by 0.2-μm Millipore filter if silver staining is to be used. If this is not carried out routinely, stain may deposit on the surface of the gel and cause staining artifacts. We have further observed that soft plastic boxes may absorb dirt or stain and produce surface spots on gels. For best results with all staining procedures, glass plates are recommended. The use of a good shaking bath with all staining procedures will keep gels from sticking to the sides of pans or to each other, a property that can result in staining irregularities.

B. Problems Associated with Silver-Staining Procedures

With more than 90 published versions of silver-staining procedures, one of the most common problems is to determine which procedure to use. We have tried many of these procedures in my laboratory and have settled on two procedures—one monochromatic and one polychromatic (see Appendix 7)—that we have found to be straightforward and reproducible. Because these methods are quite different, different problems are associated with each.

1. Fingerprints and dirt on gel surface: Because in this procedure (as most monochromatic procedures), the silver is crosslinked to proteins at the surface of the gel, any fingerprints or dirt from soiled gel plates or staining boxes will be stained. (This is not such a problem if the color-based silver stain is used, because the silver is allowed to intercalate into the protein within the gel before it is crosslinked.) It is therefore essential to wear gloves throughout this procedure and to use clean glass plates as well as glass staining pans.

2. Procedure does not work: If poor-quality reagents or old stock solutions are used for this procedure, it will not be likely to work. Because the ammonium hydroxide solution used in most of the monochromatic staining procedures will rapidly evaporate, it should be made fresh for optimal results. (This, again, is not a problem if the color-based silver stain is used.) Inadequate fixation and washing procedures will also alter results.

3. Dark or irregular backgrounds: Impurities in water or electrophoresis reagents will cause background staining by sensitive silver stains. We recommend the use of high-quality deionized water for all these procedures. Plexiglas gel electrophoresis tanks can also be used.

4. Poor color resolution: The most common reasons for inadequate color stainings are (1) poor water quality or poor reagents, (2) improper ratios of solution volume to gel volume, or (3) inadequate shaking of gels in solution. Color-coded molecular-weight standards as well as standardized staining kits are now available from

Health Products, Inc. It is advisable that these products be used when first setting up your procedure to ensure that your water quality is optimal.

C. Staining Artifacts

The most commonly observed artifacts are those described to have molecular weights of 70,000–50,000 (Tasheva and Dessev, 1983; Ochs, 1983; Marshall and Williams, 1984). These artifacts are present in most figures in this text, where silver-stained gels are shown; they can sometimes be seen as well with the less sensitive Coomassie blue stain. The artifacts are dependent on the concentration of mercaptoethanol used in the isoelectric focusing gel equilibration buffer (see discussion in Chapter 4). This staining artifact can be minimized by decreasing the concentration of β-mercaptoethanol in the equilibration buffer. Some investigators omit the mercaptoethanol completely to avoid the artifacts. This should only be done, however, if you are certain that the absence of a disulfide reducing agent will not affect your protein patterns. Other investigators have used other reducing agents such as dithiothreitol (DTT) in the equilibration buffer. While the typical mercaptan-associated artifacts are not visible, other problems with high background staining may be encountered.

VI. TROUBLESHOOTING IMMUNOBLOTTING METHODS

As with gel-staining procedures, many artifacts may be encountered with immunoblotting procedures. Most of these can be avoided if blocking, incubation, and washing procedures are optimized. A number of common problems may also occur.

A. High Background

Usually a high background is the result of inadequate blocking of binding sites on nitrocellulose or inadequate washing. It may be necessary to increase protein concentration of blocking solution or increase time of incubation with blocking solutions. Be sure that there are sufficient volumes of solutions to cover gels. Even if more dilute solutions are used, better results will be obtained if adequate shaking is used. If an excessive amount of second antibody or labeled probe is used for the final incubation, a high background may frequently be obtained. Your system should be optimized to give maximum detection of antigen (signal) without giving high background (noise) (i.e., high signal to noise ratio). It is also cost effective to determine the minimal amount of antibodies and probes that can be used for these experiments.

B. Uneven Background or Splotches on Transfer

Again, this may be due to inadequate washing and can be improved as described above. This can also be a problem if soft plastic dishes are used for multiple incubations or if nitrocellulose is handled improperly.

C. Little or No Antibody Binding to Proteins

An obvious cause of this is that the antibody does not bind to the denatured form of the antigen. Alternatively, the antibody may have a low titer. This problem can frequently be solved by affinity purification of the antibody.

D. Irregular Transfer of Protein to Nitrocellulose

This can be a problem with many gel transfer systems. It is sometimes possible to improve this by adding additional paladium wire to the chamber used for electrophoretic transfer. Another common problem is that the paper is not pressed tightly enough against the gel. This can be solved by adding additional sponges to compress the gel and paper together.

VII. CONCLUSIONS

Because there are so many different points at which problems could be encountered while carrying out high-resolution two-dimensional PAGE, initial attempts at obtaining high-quality gels may be frustrating. Once these problems are overcome, however, the results are well worth the effort. The advances in equipment that make it simple and efficient to run 10–30 reproducible two-dimensional gels have enabled us to carry out studies on a large scale to get more accurate and detailed information than is possible with one-dimensional PAGE.

REFERENCES

Anderson, N. L., and Hickman, B. J., 1979, Analytical techniques for cell fractions. XXIV. Isoelectric point standards for two-dimensional electrophoresis, *Anal. Biochem.* **93**:312–320.

Bravo, R., 1984, Two dimensional gel electrophoresis: A guide for the beginner, in: *Two Dimensional Gel Electrophoresis of Proteins: Methods and Applications* (J. E. Celis and R. Bravo, eds.), pp. 4–37, Academic Press, New York.

Chrambach, A., and Rodbard, D., 1981, "Quantitative" and preparative polyacrylamide gel electrophoresis, in: *Gel Electrophoresis of Proteins: A Practical Approach* (B. D. Hames and D. Rickwood, eds.), pp. 93–141, IRL Press, Oxford, England.

Garrels, J. I., 1979, Changes in protein synthesis during myogenesis in a clonal cell line, *Dev. Biol.* **73**:134–152.

Lane, L. C., 1982, Making electrophoresis buffers, *Trends Biochem. Sci.* **7**:242.

Marshall, T., and Williams, K. M., 1984, Artifacts associated with 2-mercaptoethanol upon high-resolution two-dimensional electrophoresis, *Anal. Biochem.* **139**:502–505.

Ochs, D., 1983, Protein contaminants of sodium dodecyl sulfate polyacrylamide gel electrophoresis, *Anal. Biochem.* **135**:470–474.

Simmen, R. C. M., Dunbar, B. S., Guerriero, V., Chafouleas, J. G., Clark, J. H., and Means, A. R., 1984, Estrogen stimulates the transient association of calmodulin and myosin light chain kinase with the chicken liver nuclear matrix, *J. Cell Biol.* **99**:588–593.

Tasheva, B., and Dessev, G., 1983, Artifacts in sodium dodecyl sulfate polyacrylamide gel electrophoresis due to 2-mercaptoethanol, *Anal. Biochem.* **129**:198–202.

Tollaksen, S. L., Edwards, J. J., and Anderson, N. G., 1981, The use of carbamylated charge standards for testing batches of ampholyte used in two-dimensional electrophoresis, *Electrophoresis* **2**:155–160.

VanBlerkom, J., 1978, Methods for the high-resolution analysis of protein synthesis: Applications to studies of early mammalian development, in: *Methods in Mammalian Reproduction* (J. C. Daniel, Jr., ed.), pp. 67–109, Academic Press, New York.

Advances in Technology of High-Resolution Two-Dimensional Polyacrylamide Gel Electrophoresis

I. INTRODUCTION

The development of methods for high-resolution two-dimensional polyacrylamide gel electrophoresis (PAGE) summarized in Chapter 1 outlines many of the advances in technology that have made possible the large-scale analysis of samples in small as well as large laboratories. This chapter is devoted specifically to the types of technologies and equipment that are available and that have been or are being developed for the improved resolution of proteins utilizing these methods.

II. EQUIPMENT AND REAGENTS FOR ISOELECTRIC FOCUSING

There are currently a variety of types of equipment for the separation of proteins by isoelectric focusing (IEF). Many of these are designed for preparative purification and utilize fluid or bead matrices such as Sephadex for suspension of ampholytes during electrofocusing. Because high-resolution two-dimensional PAGE uses polyacrylamide for the first-dimension isoelectric focusing as well as for the second-dimension separation of protein, only the equipment used for this type of focusing is discussed.

A. Isoelectric Focusing Using Horizontal Slab Apparatus

A number of horizontal gel electrophoresis apparatus systems are commercially available, such as the LKB multiphor apparatus, which is also used for immunoelectrophoresis (see Appendix 11). These systems usually include the focusing chamber for which precast slab poly-acrylamide IEF gels are available. In general, this equipment is expensive, and preparation of gels for second-dimension protein separation requires gel slicing, which is more time-consuming and generally less reproducible than the gel-tube apparatus. Additional disadvantages of slab-focusing systems have been outlined by An der Lan and Chrambach (1981). The problems encountered by this type of isoelectric focusing include (1) absorption of CO_2 across the large surface area, which could alter the pH; (2) loss of moisture from the surface, causing local increases in voltage, protein band distortion, and so forth; (3) uneven contact between electrode, resulting in an even field strength; (4) limited sample volume capacity, since sample is applied in narrow slits in gel; and (5) small electrolyte buffer reservoirs that are too small for weak electrolytes.

B. Isoelectric Focusing Using a Gel-Tube Apparatus

In general, IEF in gel tubes is preferable to horizontal slab-gel focusing for high-resolution two-dimensional PAGE for several reasons: (1) most laboratories have a gel-tube apparatus; (2) gels are easy to cast and remove from tubes for placement on second-dimension slab gel; (3) thin gels (0.7–1 mm) that yield optimal resolution of proteins by IEF can easily be prepared; (4) commercial casting and electrofocusing systems are available that make it easy to prepare and run large numbers of gels (Fig. 11.1; see also Appendix 3); and (5) the temperature of gels during focusing is more easily controlled because large volumes of electrolyte solutions surround the tubes. Tube gels can easily be cast using capillary action, as outlined in Appendix 3. These can be cast using a homemade apparatus, as shown in the Appendix 3, with commercial chambers available from Health Products (Appendix 3) or Electronucleonics (Fig. 11.1).

C. Reagents Used for Isoelectric Focusing

One of the most critical factors in achieving reproducible protein patterns using two-dimensional electrophoresis is the formation of reproducible pH gradients during IEF. Although many factors can alter the pH gradient, the quality of ampholytes is critical. Because different lots and batches of ampholytes may vary even if they are obtained from the same company, it is usually necessary to screen every batch of ampholytes. A procedure for the preparation of ampholytes using a simple inexpensive

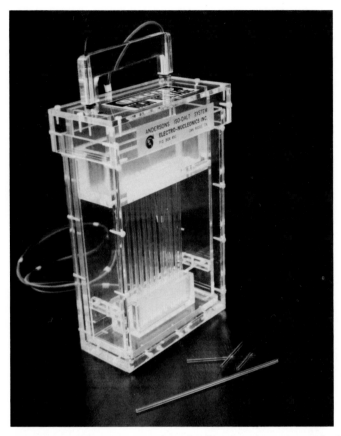

Figure 11.1. Apparatus designed to cast polyacrylamide gels and to carry out isoelectric focusing. (Photograph courtesy of Electronucleonics, Inc.)

method was recently reported (Binion and Rodkey, 1981). This procedure could provide an alternative to those who do not have ready access to commercial distributors of ampholytes. As with the use of other sources or lots of ampholytes, it is advisable to characterize your batch of ampholyte using standard proteins.

Current studies are under way to develop and prepare new types of ampholytes as well as to use ampholytes covalently attached to matrices (such as polyacrylamide). These have been designed to establish more stable ampholyte gradients (Gianazza *et al.*, 1984; Righetti *et al.*, 1983). One such product (Immobiline) is commercially available from LKB Products. The disadvantage of this product at this time is that precast gradients for horizontal focusing must be used or gradients must be formed by the investigator. Although there is great future potential for this type of product, many of the problems associated with this technology have yet to be resolved. It may not be advisable for a novice to begin analyses using this equipment.

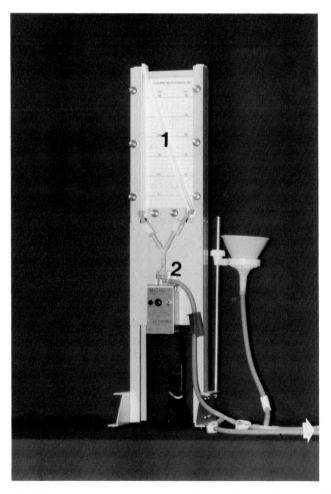

Figure 11.2. Variable-gradient maker for use in casting multiple gradient gels. The rubber spacer(s) can be reshaped to change gradient. Acrylamide is mixed in chamber 2 before entering casting chamber (arrow). (Photograph courtesy of Electronucleonics, Inc.)

III. EQUIPMENT FOR SECOND-DIMENSION POLYACRYLAMIDE SLAB GEL ELECTROPHORESIS

Electrophoresis using slab polyacrylamide gel systems such as that described by Studier (1973) (see Appendix 4) have been used routinely for many years. As discussed in Chapter 1, gradient PAGE permits the separation of a wider range of proteins than single-percentage acrylamide gels and can be used to improve the resolution of proteins.

The simple reason that most laboratories have not used gradient PAGE is that preparing individual gradient slab gels is time consuming and acrylamide gradients of gels can vary dramatically. The development of multiple slab gel casting systems, as first described by N. L. Anderson and Anderson (1978), has made it practical for the preparation of large numbers of gradient slab gels that can be reproducibly cast. The equipment that has been designed for this purpose is now commercially available from Health Products, Inc., and Electronucleonics, Inc. (see Figs. 11.2–11.4; see also methods for using this type of equipment, Appendix 4).

Figure 11.3. Diagram of chamber designed for casting multiple-gradient polyacrylamide gels. (See casting procedure outlined in Appendix 5.) (A) Rotating platform. (B) Gel-casting chamber. (C) Ports for entry of acrylamide. (Photography courtesy of Electronucleonics, Inc.)

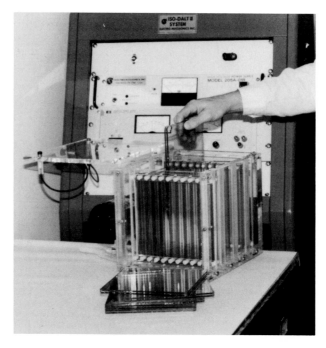

Figure 11.4. Apparatus for carrying out simultaneous electrophoresis of multiple acrylamide slab gels. (Photograph courtesy of Electronucleonics, Inc.)

Equipment for carrying out the simultaneous electrophoresis of multiple gels has also been developed (see Fig. 11.5). This equipment is currently available from Health Products, Inc., Electronucleonics, Inc., and Integrated Separation Science. This equipment has made it feasible for small laboratories as well as for centralized core facilities to process large numbers of samples for two-dimensional PAGE with minimum personnel (e.g., one person can easily run 20–40 gels per day). While a number of other gel electrophoresis systems have been described for these analyses, we have found that most of these are expensive, take up large areas of laboratory space, and are not as efficient as the equipment described above.

A. Uses of Standardized Equipment

The types of equipment specifically designed for preparation and electrophoresis of proteins are of both one and two dimensions and have greatly improved the ability of many laboratories to carry out large numbers of protein analyses with reproducibility between laboratories. (This equipment is described in detail in Section II.) This equipment has helped many laboratories standardize protein analyses that can easily be compared between laboratories. Because both IEF polyacrylamide gel and second-dimension gradient-gel lengths as well as widths are standardized, gel protein patterns are more reproducible from laboratory to laboratory.

B. Standardization of High-Resolution Two-Dimensional Polyacrylamide Gel Electrophoresis

Because of the inherent variables associated with many of the steps involved in analyzing proteins by high-resolution two-dimensional PAGE,

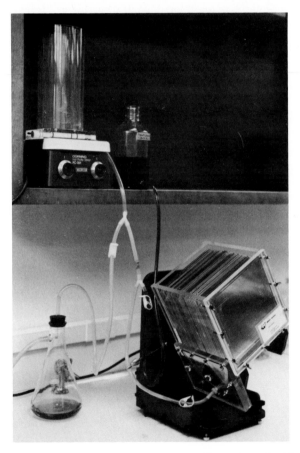

Figure 11.5. Apparatus designed for casting multiple-gradient polyacrylamide gels. (System available from Health Products, Inc., or from Integrated Separation Sciences.)

the importance of standardization in both first-dimension IEG and second-dimension SDS–PAGE cannot be emphasized enough. The major critical factors have been addressed by many laboratories that have been using these methodologies on a large scale (see reviews by Anderson and Anderson, 1981; Tracy and Anderson, 1983; Dunbar, 1985) and include (1) the use of standardized equipment for casting both IEF and polyacrylamide (gradient) gels; (2) the use of internal standards for IEF, molecular weight, and color for color-based silver-stain procedures; and (3) the presentation of two-dimensional protein patterns for publication (e.g., the orientation of cathode and anode portions of the isoelectric focusing dimension so that the pH range goes numerically from low numbers at left to higher numbers at right).

C. Specialized Equipment for Slab-Gel Electrophoresis

In addition to the equipment described above, specialized gel apparatus equipment has been designed for particular purposes.

1. *Minigel apparatus*. If investigators are carrying out studies in which very simple protein samples are being used, it is frequently advantageous to obtain results in short periods of time. For example, if one is interested in determining the change in the charge of a single or few phosphoproteins, this can be achieved using the minigel apparatus as described by Matsudaira and Burgess (1978). This system is not recommended for use with complex protein mixtures. Commercial setups for minigels are now available from companies such as BioRad (mini-vertical slab gel) or Hoefer Scientific (Mighty Small II).

2. *Giant gels*. A method has been developed that uses slab gels that are larger than most conventional gel systems (Young, 1984). While this gel system gives excellent resolution of complex radiolabeled protein samples, it is not generally practical if large numbers of samples or silver stained gels are to be analyzed.

IV. COMPUTERIZED ANALYSIS OF ONE- AND TWO-DIMENSIONAL POLYACRYLAMIDE GEL ELECTROPHORESIS

As with all areas of computerized technology, there has been a surge in both hardware and software for use with PAGE. An entire book would have to be devoted to review this area adequately, clearly beyond the scope of this text. Discussions on this topic that describe the parameters and problems to be faced in the development of computerized protein data bases for both monochromatic and polychromatic protein detection methods are given by Lipkin and Lemkin (1980), Garrels *et al.* (1984), Vincent *et al.* (1981), Sammons *et al.* (1984), Skolnick *et al.* (1982), Anderson *et al.* (1981), and Taylor *et al.* (1980, 1981, 1983). The extensive interest in these methods has led to a variety of systems, including hardware for scanning as well as analysis of two-dimensional gels. One of these systems (BioImage, Inc.) provides the software and hardware for analysis of polychromatic as well as monochromatic gels.

It is important to take into account several factors before one decides to invest the time or money in a computer-analysis system.

1. It is *not* necessary to have a computer system in order to utilize two-dimensional PAGE effectively to ask many scientific questions, as have been described in detail elsewhere in this text (e.g., antibody detection), analysis of posttranslational modification of proteins, purification of proteins, estimation of purity of proteins).

2. A computer *cannot* improve the resolution of protein analyzed by two-dimensional gel electrophoresis. In simple terms, it is better to spend time to optimize your electrophoresis techniques *before* you waste computer time analyzing gels. Frequently, it is found that if high-quality gel electrophoresis is performed, a computer may not be necessary.

3. If artifacts are present in gels (e.g., staining artifacts), the computer will have to deal with these. Investigators may therefore have to spend more time analyzing the computer analysis to determine what is real versus artifact than they would analyzing the gel itself.

4. The information obtained from computer "quantitation" is related to the method of protein detection used and therefore usually provides only a "relative" quantitation. For example, if [^{35}S]methionine is used to radiolabel proteins, the quantitation of proteins in autoradiogram will depend on the number of methionine in that protein (see discussion in Chapter 4).

If these points are seriously taken into consideration, the use of computers in data analysis and storage can be considerable. Because there are a number of commercial systems being developed in this area, these systems should become more practical, even for small laboratories. One such computer system is available from BioImage, Inc. This image analysis system is a hardware–software intact system that is user friendly; it has been designed to analyze polychromatic as well as monochromatic gels.

V. COMMERCIAL AVAILABILITY OF PRECAST GRADIENT GELS AND GEL-ANALYSIS SERVICES

Many laboratories have had difficulty preparing reproducible gradient gels, and some laboratories need only to run an occasional two-dimensional gel for protein analysis. For this reason, it may be advantageous for a laboratory to purchase precast gradient gels for use with electrophoresis chambers that are now available from commercial sources (e.g., Integrated Separation Systems).

Alternatively, it is anticipated that it will be possible to have your sample analyzed by service companies. If limited numbers of analyses are needed, this is a practical approach. We have found, however, that most studies require the analysis of many samples, including replicates as well as multiple protein loads. The cost of these analysis services is therefore generally out of the realm of practicality for most research grants.

REFERENCES

An der Lan, B., and Chrambach, A., 1981, Analytical and preparative gel electrofocusing, in: *Gel Electrophoresis of Proteins: A Practical Approach* (B. D. Hames and D. Rickwood, eds.), pp. 157–188, IRL Press, Oxford, England.

Anderson, N. L., and Anderson, N. G., 1978, Analytical techniques for cell fractions. XXII. Two-dimensional analysis of serum proteins and tissue proteins: Multiple gradient-slab electrophoresis, *Anal. Biochem.* **85**:341–354.

Anderson, N. L., Taylor, J., Scandora, A. E., Coulter, B. P., and Anderson, N. G., 1981, The TYCHO system for computerized analysis of two-dimensional gel protein mapping data, *Clin. Chem.* **27**:1807–1820.

Binion, S., and Rodkey, L. Scott, S., 1981, Simplified method for synthesizing ampholytes suitable for use in isoelectric focusing of immunoglobulins in agarose gels, *Anal. Biochem.* **112**:362–366.

Dunbar, B. S., 1985, Protein analysis by high-resolution two-dimensional polyacrylamide gel electrophoresis, in: *Methods in Molecular Endocrinology* (W. Schrader and B. W. O'Malley, eds.), pp. 000–000, Houston Biological Association, Houston, Texas.

Garrels, J. I., Farrar, J. T., and Burwell, C. B., 1984, The QUEST system for computer analyzed two-dimensional electrophoresis of proteins, in: *Two-Dimensional Gel Electrophoresis of Proteins: Methods and Applications* (J. Celis and R. Bravo, eds.), pp. 38–92, Academic Press, New York.

Gianazza, E., Frigerio, A., Tagliabue, A., and Righetti, P. G., 1984, Serum fractionation on immobilized pH gradients with one- and two-dimensional techniques, *Electrophoresis* **5**:209–216.

Lipkin, L. E., and Lemkin, P. F., 1980, Data-base techniques for multiple two-dimensional polyacrylamide gel electrophoresis analyses, *Clin. Chem.* **26**:1403–1412.

Matsudaira, P. T., and Burgess, D. R., 1978, SDS microslab linear gradient polyacrylamide gel electrophoresis, *Anal. Biochem.* **87**(2):386–396.

Righetti, P. G., Gianazza, E., and Bjellquist, B., 1983, Modern aspects of isoelectric focusing: Two-dimensional maps and immobilized pH gradients, *J. Biochem. Biophys. Methods* **8**:89–108.

Sammons, D. W., Adams, L. D., Vidmar, T. J., Hatfield, C. A., Jones, D. H., Chuba, P. J., and Crooks, S. W., 1984, Applicability of color silver stain (Gelcode System) to protein mapping with two-dimensional gel electrophoresis, in: *Two-Dimensional Electrophoresis of Proteins: Methods and Applications* (J. Celis and R. Bravo, eds.), pp. 112–127, Academic Press, New York.

Skolnick, M. M., Sternberg, S. R., and Neel, J. V., 1982, Computer programs for adapting two-dimensional gels to the study of mutation, *Clin. Chem.* **28**:969–978.

Studier, W. F., 1973, Analysis of bacteriophage T7 early RNAs and proteins on slab gels, *J. Mol. Biol.* **79**:237–248.

Taylor, J., Anderson, N. L., Coulter, B. P., Scandora, A. E., and Anderson, N. G., 1980, Estimation of two-dimensional electrophoretic spot intensities and positions by modeling, in: *Electrophoresis '79* (B. Radola, ed.), pp. 329–339, Walter deGruyter, Berlin.

Taylor, J., Anderson, N. L., and Anderson, N. G., 1981, A computerized system for matching and stretching two-dimensional gel patterns represented by parameter lists, in: *Electrophoresis '81* (R. Allen and P. Arnaud, eds.), pp. 383–400, Walter deGruyter, Berlin.

Taylor, J., Anderson, N. L., and Anderson, N. G., 1983, Numerical measures of 2-D gel resolution and positional reproducibility, *Electrophoresis* **3**:338–346.

Tracy, R. P., and Anderson, N. L., 1983, Applications of two-dimensional gel electrophoresis in the clinical laboratory, in: *Clinical Laboratory Annual* (H. A. Hombruger and J. G. Butsabs, eds.), pp. 101–130, Appleton-Century-Crofts, East Norwalk, Connecticut.

Young, D. A., 1984, Advantages of separations on "giant" two-dimensional gels for detection of physiologically relevant changes in the expression of protein gene products, *Clin. Chem.* **30**(12, pt. 1):2104–2108.

Appendixes

Appendix 1

Protein Assays

I. LOWRY PROTEIN ASSAY

Lowry, O. H., Rosebrough, N. J., Farr, A. L. and Randall, R. J., 1951, Protein measurement
with the folin phenol reagent, *J. Biol. Chem.* **193**:265–275.

A. Reagents

1. Sodium carbonate ($Na_2CO_3 \cdot H_2O$) (Sigma)*
2. Sodium hydroxide pellets (NaOH) (Sigma)
3. Copper sulfate ($CuSO_4 \cdot 5H_2O$) (Mallinckrodt)
4. Sodium potassium tartrate (NaK tartrate·$4H_2O$) (Sigma)
5. Bovine serum albumin (BSA) (Sigma)
6. Folin-Ciocalteu's phenol reagent, 2.0 N (Sigma)

B. Stock Solutions

1. Solution A
   a. 2% Na_2CO_3, 2 g
   b. 0.1 N NaOH, 0.4 g
   c. Bring to 100 ml with distilled H_2O; store stock in dark bottle
2. Solution B_1
   a. 1% $CuSO_4 \cdot 5H_2O$, 0.5 g
   b. Bring to 50 ml with distilled H_2O
3. Solution B_2
   a. 2% NaK tartrate·$4H_2O$, 1 g
   b. Bring to 50 ml with distilled H_2O

*Manufacturers in parentheses.

    4. Solution C
       a. 100 part A
       b. 1 part B_1
       c. 1 part B_2
    5. BSA Standard
       a. BSA 100 mg
       b. Bring to 100 ml with distilled H_2O. Always read UV absorbance
          of standard on spectrophotometer at 280 nm. The absorbance
          of BSA at 1 mg/ml is 0.58. If your standard stock solution is
          off, you should adjust it accordingly or use calculations to
          adjust your values. Freeze small (1.0-ml) aliquots and thaw just
          before use. BSA is commonly used as a protein standard, al-
          though other proteins with different amino acid compositions
          such as lysozyme may also be used.

C. Procedure

    1. Prepare protein standards, each in duplicate or triplicate: pipette
       aliquots of BSA stock into 12 × 75 mm tubes (5–40 μl will give
       absorbance readings in the linear range of the assay). Add distilled
       H_2O to bring volumes to 200 μl. Be sure to include a reagent blank
       of 200 μl H_2O and all other reagents, but no BSA.
    2. Prepare protein samples, also in duplicate or triplicate: Pipette an
       aliquot of sample containing 5–30 μg of protein into tube. (If
       protein concentration is completely unknown, try 10-, 50-, and
       100-μl aliquots.) Bring volumes to 200 μl with distilled H_2O.
    3. Add 1.0 ml solution C to each tube; mix and incubate 10 min at
       room temperature.
    4. Add 50 μl Folin-Ciocalteu's reagent and vortex immediately.
    5. Incubate for at least 30 min at room temperature.
    6. Read absorbance at 660 or 750 nm with spectrophotometer (tung-
       sten light setting). Blank against appropriate reagent blank (200 μl
       H_2O + 1.0 ml solution C + 50 μl Folin-Ciocalteu's reagent). Plot
       amount of BSA in standards against average absorbance for each,
       and use this standard curve to determine amount of protein in
       samples.

II. BRADFORD COOMASSIE BLUE ASSAY

Bradford, M., 1976, A rapid and sensitive method for the quantitation of microgram quan-
    tities of protein utilizing the principles of protein–dye binding, *Anal. Biochem.* **72**:248–254.

This assay is used to determine protein concentrations within the range of 150–1500 μg/ml.

A. Reagents*

1. Coomassie Brilliant Blue G-250 (BioRad or Sigma)
2. Mercaptoethanol (BioRad or Sigma)
3. 95% ethanol
4. Phosphoric acid (H_3PO_4), 85% (Mallinckrodt)

B. Stock Solutions

1. Solution A
   a. Coomassie Brilliant Blue G-250, 100 mg
   b. Bring to 50 ml with 95% ethanol
2. Solution B
   a. 8.5% phosphoric acid, 100 ml, 85%
   b. 50 ml solution A
   c. 850 ml distilled H_2O
   d. Allow to stir 3–4 hr or overnight before filtering stock solution with Whatman filter paper

C. Procedure for Standard Coomassie Protein Assay

1. Prepare BSA standard as described for the Lowry protein assay. The same buffer used for protein samples should be used to bring the final volumes to 100 μl.
2. Pipette 10-, 20-, 50-, and 100-μl samples into 12 × 75 mm test tubes and adjust protein sample volume to 100 μl with appropriate buffer or distilled H_2O.
3. Add 5 ml Coomassie blue stock reagent (solution B) to every sample while vortexing.
4. After 2 min (before 1 hr), read absorbance at 595 nm. Blank against appropriate buffer blank (100 μl buffer plus 5 ml Coomassie reagent). Plot amount of BSA in standards against average absorbance for each, and use standard curve to determine protein in unknown samples.

*All standardized reagents can be obtained as kit from BioRad.

D. Procedure for Microprotein Coomassie Assay

For determining protein concentrations in the range of 1–25 μg/ml.

1. Prepare BSA standards as described for standard Coomassie assay, 1–10 μg per tube.
2. Pipette 10-, 20-, 50-, and 100-μl samples into 12 × 75 mm test tubes. Adjust final volumes to 100 μl with appropriate buffer or distilled H_2O.
3. Add 1 ml Coomassie blue stock (solution B) and vortex.
4. Read absorbance at 595 nm after 5 min. Plot standard curve and determine protein in unknown samples.

III. MODIFIED BRADFORD PROTEIN ASSAY FOR SAMPLES IN UREA SOLUBILIZATION BUFFER

Ramagli, L. S., and Rodriguez, L. V., (1985) Quantitation of microgram amounts of protein in two-dimensional polyacrylamide gel electrophoresis sample buffer, *Electrophoresis* **6**:559–563.

This is an excellent method for quickly determining the amount of protein in a sample that has been solubilized in urea gel electrophoresis solubilization buffer. It is often the case that the amount of protein in a sample cannot be assayed before solubilization. For example, cells and sometimes tissues are put directly into solubilization buffer, or the buffer is required to obtain the sample off an inert support (see Appendix 2, Section III). The use of ovalbumin is the standard for this assay, but the investigator should select the standard protein that responds similarly to most of the proteins in the sample (i.e., BSA for serum proteins).

A. Reagents

1. Protein for standard (e.g., BSA, ovalbumin)
2. Coomassie Brilliant Blue G-250 dye reagent (BioRad, Pierce)

B. Stock Solutions

1. Urea electrophoresis solubilization buffer (see Appendix 2, Section I.C.2)

2. Standard protein stock (BSA, ovalbumin, or protein standard of your choice)
   a. 0.5% protein standard, 5 mg
   b. Bring to 1 ml with urea buffer
3. 0.1 N HCl
   a. concentrated HCl, 86 μl
   b. Bring to 10 ml with distilled H_2O
4. Coomassie dye reagent
   Can be made according to Bradford (1976) as described in Appendix 1 or bought from Pierce (ready to use: 23200) or BioRad (concentrate that must be diluted and filtered before use: 500-0006).

C. Procedure

1. Prepare protein standard samples in duplicate to contain 1–50 μg protein in 10 μl urea buffer.
2. Add 10 μl 0.1 N HCl to each tube and bring the volume up to 100 μl with 80 μl distilled water.
3. Prepare unknown samples similarly. For example, bring 5–7 μl of a protein sample solubilized in urea buffer up to 10 μl with more urea buffer, add 10 μl 0.1 N HCl and bring to 100 μl with 80 μl distilled water.
4. Add 3.5 ml Coomassie dye reagent to each tube, mix thoroughly.
5. After 5 min read the absorbance at 595 nm with spectrophotometer.
6. Use a blank containing all reagents except protein to set the zero.

REFERENCES FOR METHODS FOR ADDITIONAL PROTEIN ASSAYS

Modifications of Lowry Procedure

Esen, A., 1978, A simple method for quantitative, semiquantitative and qualitative assay of protein, *Anal. Biochem.* **89:**264–273.
Peterson, G. L., 1977, A simplification of the protein assay method of Lowry *et al.* which is more generally applicable, *Anal. Biochem.* **83:**346–356.

Fluorescence Protein Assays

Castell, J. V., Cervera, M., and Marco, R., 1979, A convenient micromethod for the assay of primary amines and protein with fluorescanine. A reexamination of the conditions of reaction, *Anal. Biochem.* **99:**379–391.

Flores, R., 1978, A rapid and reproducible assay for quantitative estimation of proteins using bromophenol blue, *Anal. Biochem.* **89:**605–611.

Johnson, M. K., 1978, Variable sensitivity in the microassay of proteins, *Anal. Biochem.* **86:**320–323.

Robrish, S. A., Kemp, C., and Bowen, W. H., 1978, *Anal. Biochem.* **84:**196–204.

Udenfriend, S., Stein, S., Böhlen, P., Dairman, W., Leimgruber, W., and Weigele, M., 1972, Applications of fluorescence: A new reagent for assay of amino acids, peptides, proteins and other primary amines in the picomole range, *Science* **178:**871.

Ultraviolet Absorption Assays

Peterson, G. L., 1983, Determination of total protein, *Methods Enzymol.* **91:**95–119.

Warburg, O., and Christian, W., 1941, Isolierung und Kristallisation des Garungsterments Enolase, *Biochem. Z.* **310:**384–421.

Sample Preparation and Solubilization for One-Dimensional Polyacrylamide Gel Electrophoresis and Isoelectric Focusing*

I. SAMPLE SOLUBILIZATION FOR ISOELECTRIC FOCUSING

A. Reagents

1. Sodium dodecyl sulfate (SDS) (BioRad)†
2. Cyclohexylaminoethane (CHES) (Calbiochem)
3. Glycerol (Fisher)
4. β-Mercaptoethanol (BioRad)
5. Urea (Ultrapure) (BioRad)
6. Nonidet P-40 (nonionic detergent) (Accurate Chemical)
7. Ampholytes (pH 3.5–10: BioRad, LKB, or Pharmacia; pH 2–11: Serva). This wide range mixture is adequate for most routine samples. Other pH range or combinations of brands of ampholytes may be used in some instances (see discussions in Chapters 1 and 10).
8. H_2O, deionized with mixed bed resin (Continental Filter System) or deionized double-distilled H_2O
9. Celite (Sigma, registered trademark for diatomaceous silica products)

B. Equipment and Supplies

1. 1–3-ml glass vials with screwtops
2. 20- and 100-μl Hamilton syringes

---

*Modified from Anderson, N. G., Anderson, N. L., and Tollaksen, S. L., 1983, Operation of the ISO-DALT System, ANL-BIM, 84-1.
†Manufacturers in parentheses.

3. Ultracentrifuge (for removing insoluble material and nucleic acids from samples)
   a. Beckman Rotor Ti-42.2 (Fig. A2.1; this rotor is convenient for small samples, but other angular rotors can also be used)
   b. Beckman Airfuge (can be used for SDS solubilization but is not advised for urea solubilization, as heating samples can cause carbamylation of proteins)
4. Amicon centrifugation concentrator tubes (Centricon 10, #4205) (optional but very useful for concentrating samples without increasing salt concentration)
5. 0.2-μm syringe filters (Nalgene Labware 190-2020; Nalge Co., Sybron Corp.)
6. Microdialyzer (Health Products)

C. Stock Solutions

1. SDS solubilization buffer for isoelectric focusing (IEF)
   a. 0.05 M CHES, 100 mg
   b. 2% SDS, 200 mg
   c. 10% glycerol, 1 ml

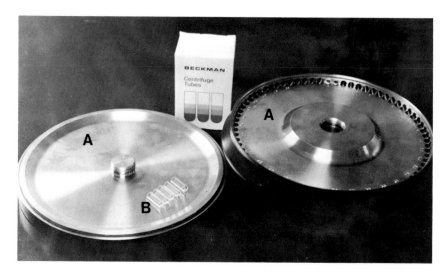

Figure A2.1. Rotor (Beckman Ti-42.2) (A) and tubes (B) optimally used for ultracentrifugation of small volume samples (20–200 μl) to remove nucleic acids on insoluble material prior to isoelectric focusing.

    d. Bring to 10 ml with H_2O and pH to 9.5. Add a very small amount of bromophenol blue.

    e. Filter with 0.2-μm syringe filter.

    f. Add 0.2 ml (2%) β-mercaptoethanol just prior to use.

2. Urea solubilization buffer for IEF

    a. 9 M urea, 54 g

    b. (4%) Nonidet P-40, 4 ml

    c. Add water to final volume of 100 ml and filter with a 0.2-μm filter.

    d. These reagents can be frozen in small aliquots, except that the β-mercaptoethanol should be added immediately before use.

    e. Add 2% β-mercaptoethanol and 2% ampholytes just prior to use to the small aliquot of solubilization buffer. (Use either 3.5-10 LKB or 9-11 Serva ampholytes.)

D. Procedure

1. Choice of Solubilization Method

We have found that the following two solubilization recipes described by Tollaksen *et al.* (1984)* as described in Sections C.1 and C.2 can be used for most protein samples. You may have to try each the first time to determine the method and concentrations to be used in order to optimize solubilization of some samples. Different solubilization methods may affect the apparent pI of some proteins. It is critical that the ratio of solubilization buffer to sample be optimized (see discussions in Chapters 3 and 10).

Although you will have to optimize the solubilization method for your exact samples, in general, we have found the following ratios of protein to sample buffer adequate:

    200–500 μg tissue homogenate per 2 ml solubilization buffer

    20–50 μl cell pellet per 300 μl solubilization buffer

    1 $\times$ 10^6 cells in tissue culture plate; flood plate with 500 μl solubilization buffer to solubilize cells directly

    10–200 μg soluble protein/30–50 μl solubilization buffer

Note: 5–30 μl of each of the above samples should be adequate for identification of abundant proteins by Coomassie blue or of minor proteins by silver stain in two-dimensional gel patterns.

*Tollaksen, S. L., Anderson, N. L., and Anderson, N. G., 1984, *Operation of the ISO-DALT System,* 7th ed., p. 1, ANC-BIM-84-1.

2. Preparation of Sample Prior to Solubilization

Most protein samples can be solubilized directly. If tissues are used, they may be homogenized in an ice bath. For better solubilization of tissues, it is advisable to use tissues frozen in liquid nitrogen. These can be pulverized in liquid nitrogen, as shown in Figures A2.2 and A2.3. The pulverized tissue can be solubilized directly or extracted with water and centrifuged to remove insoluble from water-soluble proteins (see discussion in Chapter 3).

3. Sample Concentration

If necessary, concentrate protein sample using Amicon microconcentrator tubes or lyophilize sample directly or following dialysis against ammonium bicarbonate (pH 7.5) or H_2O to remove salts. Alternatively, a two-times concentrated sample buffer can be used so that sample will not be diluted as much.

4. Sample Solubilization

a. Urea Solubilization. Samples should be suspended in the urea solubilization solution and incubated at room temperature for 2 hr. *Do not heat, or you will generage charge artifacts.*

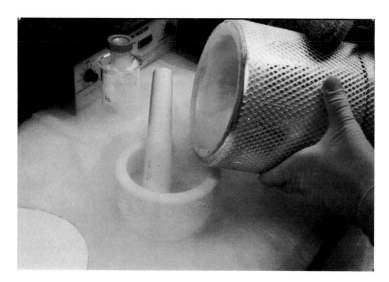

Figure A2.2. Preparation of tissue homogenates. Liquid nitrogen is poured into mortar containing frozen tissue.

Figure A2.3. Preparation of tissue homogenates. Sample is pulverized in liquid nitrogen with a pestle until it is a uniform powder. This method of tissue preparation is generally better than tissue homogenization because proteolysis is minimized.

b. SDS Solubilization. Samples should be suspended in SDS solubilization buffer and placed in a tightly capped glass vial and heated for 5–10 min in a boiling waterbath. (Thick plastic tubes such as microfuge tubes are insulated and interfere with heating.) It may be necessary to solubilize some samples at room temperature for 2–3 hr with or without heating.

5. Centrifugation

Following the incubation, samples are centrifuged to remove nonsolubilized material and nucleic acids that may interfere with focusing or cause streaking in second-dimension protein patterns (100,000–200,000 × g for 2 hr is suggested). We recommend using a Beckman Ti-42.2 rotor, which holds 72 tubes (Fig. A2.1). A main advantage of this rotor is that small sample volumes (20–200 μl) can be prepared in these tubes (Beckman 342.303). A problem encountered occasionally is that some proteins of interest may be removed during centrifugation if they are not adequately solubilized. Centrifugation should be done at room temperature to prevent either urea or SDS from precipitating out of the solubilization buffer.

6. Loading the Sample

Following centrifugation, supernates should be removed from the tube and applied to IEF gels. While some samples can be frozen at $-70°C$ and run at a later date, best results are obtained if samples are run immediately after solubilization.

II. SAMPLE PREPARATION FOR ONE-DIMENSIONAL SODIUM DODECYL SULFATE POLYACRYLAMIDE GEL ELECTROPHORESIS

Laemmli, U. K., 1970, Cleavage of structural proteins during the assembly of the head of bacteriophage T4, *Nature (Lond.)* **277**:680–685.

A. Reagents

1. Sodium dodecyl sulfate (BioRad)
2. Trizma Base (Sigma)
3. β-Mercaptoethanol (BioRad)
4. Glycerol (Fisher)

B. Equipment and Supplies

Same as in Section I.

C. Stock Solution: SDS Solubilization Buffer (pH 6.8)

1. 2% SDS, 2 g
2. 0.0625 M Trizma Base, 0.75 g
3. 10% glycerol, 10 ml; fill to 100 ml.
4. 2–5% β-mercaptoethanol; add fresh to small aliquot.

D. Procedure

1. Add solubilization buffer to sample.
2. Heat at 95°C in boiling water bath. (It may be adequate and some-

times advantageous to incubate sample for 2–3 hr at 25°C if proteases are not present in sample.)
3. Ultracentrifuge at 100–200,000 $\times$ g to remove nucleic acids or insoluble material if necessary.

III. CONCENTRATION OF PROTEINS FOR GEL ELECTROPHORESIS FROM CONDITIONED MEDIUM USING DIATOMACEOUS EARTH

Maresh and Dunbar (manuscript in preparation).

This procedure is used to concentrate medium conditioned by tissue culture cells in order to study secretory proteins by two-dimensional PAGE. It is an excellent method with which to compare serum-free medium from cell cultures under various conditions (the ubiquitous serum proteins make it difficult to analyze differences in gel patterns). Large numbers of small quantity media samples can be concentrated quickly and inexpensively for gel electrophoresis without the resultant artifacts generated by the high concentration of salts in samples concentrated by other methods. Initial studies should be done on a particular system to optimize the amount of diatomaceous earth needed, its length of incubation with the medium, and how much urea buffer will be used. The following procedure has been optimized for our system and should be used as a guideline.

A. Reagents

Diatomaceous earth, 97.5% SiO_2 (Sigma, grade III #D-5384) (grade III has been found to be most useful for this method of the three grades offered by Sigma).

B. Equipment

Microcentrifuge (Beckman B microfuge, Fisher model 235B).

C. Stock Solutions

1. 10% diatomaceous earth
   a. Diatomaceous earth, 1 g

    b. Wash by adding at least 20 ml distilled water, shaking, centrifuging the diatomaceous earth into a pellet, and removing the water.

    c. Repeat washing. Add 10 ml distilled water and use this as a 10% solution.

  2. Urea gel electrophoresis solubilization buffer (see Section I.C.2)

D. Procedure

1. Remove cellular debris from medium conditioned by cells in culture by centrifugation. Add the diatomaceous earth solution (making sure it is well suspended) to the conditioned medium in a plastic tube to make a final concentration of 2% diatomaceous earth solution (i.e., 200 μl 10% diatomaceous earth added to 800 μl conditioned medium). Mix gently and continuously by using a rocker platform or rotator for 2 hr at room temperature.

2. Pellet the diatomaceous earth by centrifugation. A quick way to do this is to put the mixture into Eppendorf tubes and centrifuge. Remove the supernatant. Add urea gel electrophoresis solubilization buffer to the pellet; the amount will depend on how much protein was in the medium and how much sample can be loaded onto the gels (we use 30–70 μl of urea buffer per original ml of serum-free medium). Let stand at room temperature for 2 hr vortexing every 15 min. Pellet the diatomaceous earth and load supernatant onto a gel.

IV. PREPARATION OF ISOELECTRIC POINT STANDARDS FOR TWO-DIMENSIONAL ELECTROPHORESIS

N. L. Anderson and B. J. Hickman, 1979, Analytical techniques for cell fractions. XXIV. Isoelectric point standards for two-dimensional electrophoresis, *Anal. Biochem.* **93**:312–320.

A. Reagents

1. Glyceraldehyde 3-phosphate dehydrogenase (GPD) EC 1.2.1.12 (Sigma G0763)
2. Ammonium bicarbonate (Sigma)
3. Urea (BioRad)

B. Equipment and Supplies

    1. Water bath
    2. Dialysis tubing

C. Stock Solutions

    1. 8 M urea
       a. Urea, 4.8 g
       b. Bring to 10 ml with distilled H_2O
    2. 0.1 M NH_4CO_3
       a. NH_4CO_3, 7.9 g
       b. Bring to 1 liter with distilled H_2O

D. Procedure

    1. Dialyze the GPD extensively against 0.1 M ammonium bicarbonate to remove ammonium sulfate; Sigma GPD is sent as a crystalline suspension in 2.6 M $(NH_4)_2SO_4$.
    2. Read the A276, multiply by 0.95 to determine milligrams per milliliter of the dialyzed sample (Decker, 1977, p. 43).*
    3. Lyophilize approximate 5-mg aliquots.
    4. To begin carbamylation process, add 1 ml 8 M urea and place the preparation in a 95°C water bath.
    5. Remove 0.1-ml samples at 2, 4, 8, 10, 15, 20, 30, and 45 min; cool quickly in an ice bath.
    6. Alternatively, the preparation can be aliquoted at zero time and removed from the water bath at the respective intervals.
    7. Samples of each time point will have to be analyzed by IEF or two-dimensional PAGE. The samples should be pooled to give charge train comparable to that shown in Chapter 3, Fig. 3.3.

*Decker, Lillian (ed.), 1977, *Worthington Enzyme Manual,* pp. 42–43, Worthington Biochemical Corporation, Freehold, New Jersey.

Appendix 3

Method for Isoelectric Focusing

I. CASTING GELS USING INDIVIDUAL TUBES

A. Reagents

1. Urea, ultrapure (BioRad)*
2. Ampholytes (LKB, Serva, or Pharmacia recommended); pH will depend on needs of investigator. For general use, we mix 1 part Pharmacia 3–10 with 2 parts LKB, pH 3.5–10, to avoid pH gaps).
3. Acrylamide (BioRad)
4. Bis-acrylamide (BioRad)
5. Nonidet P-40 (Accurate)
6. Ammonium persulfate (BioRad)
7. TEMED (BioRad)
8. Sodium hydroxide (Sigma)
9. Phosphoric acid (Fisher)
10. Chromerge (Fisher)

B. Equipment

1. Glass Tubes for Gels

Although a variety of tube sizes can be used for isoelectric focusing (IEF), the size will depend on the type of experiment. For optimal resolution of complex protein systems, use the size tubes described by Anderson and Anderson (1978).† These can be made easily and cheaply by cutting off the tips of 0.2-ml disposable glass pipettes available from most

---

*Manufacturers in parentheses.

†Anderson, N. G., and Anderson, N. C., 1978, Analytical techniques for cell fractions. XXI. Two-dimensional analysis of serum and tissue proteins: Multiple isoelectric focusing, *Anal. Biochem.* **85:**331–340.

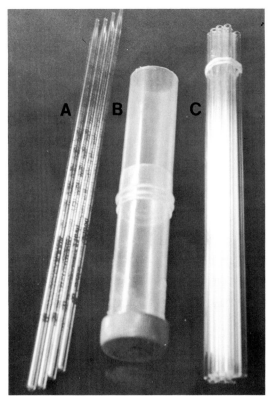

Figure A3.1. Supplies needed to cast multiple isoelectric focusing gels rapidly using capillary action. (A) 0.2-ml glass serological pipettes. (B) Two 50-ml conical plastic centrifuge tubes. Note that "cones" of tubes have been cut off and that one tube has been tightly fitted over the one with the cap to extend the length of the tube. (C) Serological pipettes that have had their tips cut off and numbering removed by chromic acid.

general scientific suppliers (American Scientific Products, S/P disposable serological pipettes, 0.2 ml, #P4644-2T) (see Fig. A3.1). This size tube has several advantages: (1) excellent resolution in the IEF dimension can be obtained; (2) the gels are easily removed from the tubes without treating the tubes with silicone, and so forth; and (3) the gels will not fall out of the tubes during electrophoresis, as sometimes happens when larger tubes are used. The tips of the pipettes can be cut off, and the markings are removed when the tubes are soaked in Chromerge. Isoelectric focusing tubes are prepared by soaking in Chromerge for several hours or by heating for 1 hr. (Use caution when working with this solution.) Rinse the tubes thoroughly with distilled or Continental filtered H_2O and dry thoroughly.

2. Gel-Casting Apparatus

The apparatus consists of two 50-ml conical centrifuge tubes with one lid (see diagram in Fig. A3.1).

3. Gel Electrophoresis Apparatus

Most any tube gel electrophoresis apparatus can be used if appropriate grommets or corks are prepared to fit small tubes (e.g., BioRad electrophoresis unit model 175 tube gel apparatus). This unit can also be used for electroelution, as illustrated in Appendix 13 (Fig. A13.1).

4. Power Supply

Most any high-voltage power supply can be used, but high amperage capacity is needed if one power supply is to be used for both first- and second-dimension gels (minimum 700-V capacity). We recommend Health Products power supply as adequate for both first- and second-dimension electrophoresis. This power supply is designed to shut off if upper gel buffer leaks into lower tank, so that your electrophoresis chamber will not be destroyed if gels fall out of tubes during IEF.

C. Stock Solutions

1. Isoacrylamide stock
   a. 30% acrylamide, 30 g
   b. 1.8% bis-acrylamide, 1.8 g
   c. Bring to 100 ml with distilled H_2O
   d. Filter to 0.2 μm
2. Electrode buffers for isoelectric focusing
   a. Upper electrode buffer (0.02 N NaOH): The upper electrode buffer is prepared by degassing H_2O just before adding to the electrode chamber. The volume will depend on the apparatus used. Place 200 ml H_2O in a 600-ml Virtis lyophilization flask and degas using the lyophilizer. Take great care not to let solutions "boil" too quickly, to avoid getting moisture in lyophilizer. (Again, we found the laboratory vacuum line time consuming and inadequate for thorough degassing.) 0.4 ml of 10 N NaOH is added to the water *after degassing,* to avoid Na_2CO_3 formation.
   b. Lower electrode buffer: The bottom chamber is filled with 0.085% phosphoric acid. It is not necessary to degas this solution.
3. Second-dimension equilibration buffer
   a. 0.125 M Tris base, 3 g in 150 ml pH 6.8
   b. 2% SDS, 4 g
   c. 10% glycerol, 20 ml; fill to 200 ml

   d. Very small amount of bromophenol blue
   e. Filter to 0.2 μm
   f. Freeze 10- to 15-ml aliquots
   g. Add 0.5–0.8% β-mercaptoethanol to aliquots just before using

D. Procedure for Casting IEF Gels

1. To cast 20–22 IEF gels having the dimension described in Section I.C.1:
   a. Urea, 8.25 g
   b. Ampholytes, 0.75 ml [of desired range, usually pH 3.5–10 mixtures of several suppliers (LKB or Pharmacia)]
   c. Isoacrylamide stock, 2.0 ml
   d. Distilled H_2O, 6.0 ml (degas after mixing these reagents)
   e. Nonidet P-40, 0.3 ml
   f. 10% ammonium persulfate, 70 μl
   g. N,N,N′,N′-tetramethylethylenediamine (TEMED), 10 μl
2. The urea is dissolved in the H_2O by swirling the flask under warm running water (or in a microwave oven for approximately 5 sec). *Do not heat solution.*
3. The acrylamide and ampholytes are added to the mixture of water and urea, and the solution is degassed on a lyophilizer. (We have

Figure A3.2. Lyophilization flask used to degas urea–acrylamide–ampholyte mixture. Urea may crystallize when extensively degassed. This flask can be warmed carefully to bring the mixture back into solution. *Do not heat solution.*

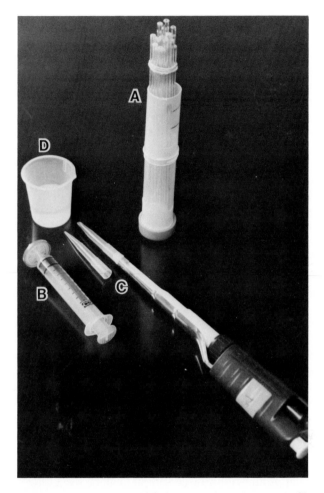

Figure A3.3. Equipment used to cast multiple isoelectric focusing gels. (A) Glass tubes and plastic casting chamber (described in Figure A3.1). (B) Syringe with 0.2 μm Millipore filter. (C) 1-ml pipettor with pipette tip cut off to enlarge opening for detergent pipetting. (D) 30-ml beaker.

tried using a laboratory vacuum line and have found that this is frequently not adequate.) Since the urea may come out of the solution as you degas (Fig. A3.2), you can warm it back to room temperature and gently swirl the flask until it goes back into solution.

4. Add the Nonidet P-40 and stir slowly to mix. [*Hint:* A large, plastic Eppendorf pipette tip can be cut off for easier and more accurate pipetting of viscous detergents) (see Fig. A3.3).] It is recom-

mended that this mixture be filtered with a 0.2-μm filter fitted to
a syringe (Fig. A3.3).

5. Use a rubber band to hold the glass tubes together (Fig. A3.3).
6. Add the catalyst to the acrylamide solution, swirl the flask gently
   to mix, and pour into the plastic tube apparatus. Lower the glass
   tubes into the acrylamide, and gently but quickly fill the plastic
   tube with H_2O using the wash bottle (Fig. A3.4). You should be
   able to see the acrylamide level clearly in the tubes. We routinely
   measure the distance of the rubber band and use this as a marker
   for acrylamide height. Allow the gels to polymerize for at least
   1 hr.
7. Rinse the gels with H_2O and cut the excess acrylamide from the
   bottom of the tubes. This is accomplished by removing the plastic
   tube cap and rotating the group of glass tubes on a piece of parafilm. The gels are then placed into the electrophoresis apparatus.
   (If Commercial Casting Systems are used, the gels are cast in the
   chamber so that they do not have to be handled at this point.)
8. Prefocus the gels at 200 V for 1–2 hr. In theory, this will remove
   the sulfate ions from the ammonium persulfate as well as other
   nonamphoteric ions, which may interfere with the focusing. We
   have frequently omitted this step, however, with no noticeable
   differences in protein patterns.
9. Load the protein samples. A Hamilton syringe can be used if care
   is taken to wash it thoroughly between sample application. Up to
   30 μl of sample should contain a maximum of 200 μg total protein
   of a complete mixture, or less for more purified proteins if silver
   stain is to be used. It may contain up to 0.5 mg for Coomassie
   blue staining. A larger volume is not generally acceptable if 0.2-
   ml pipette tubes are used. For some preparative runs, however,
   a larger volume (70–100 μl sample) can be used on gels 1.5–2 mm
   in diameter. *Make sure to remove all air bubbles from the top and
   bottom of the gels.*

E. Procedure for Isoelectric Focusing

1. In general, samples are focused for 10,000–12,000 V-hr (e.g., 17
   hr at 700 V). This will depend on the nature of your sample (see
   above discussion on parameters affecting IEF). We have found
   that you will usually get better resolution of proteins if you focus
   for a shorter period of time at higher voltage (i.e., 700 V for 16
   hr is better than 500 V for 22 hr).

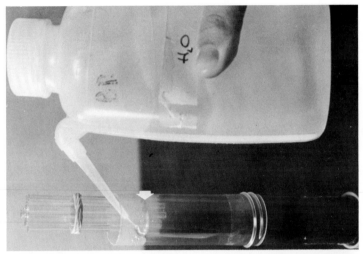

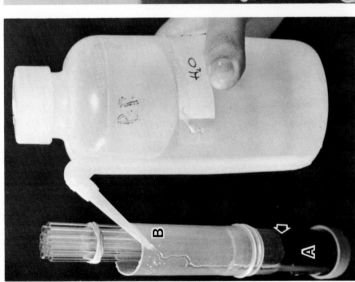

Figure A3.4. Casting isoelectric focusing gels. (Left) Acrylamide mixture (A) is placed in casting tube to approximate level shown by arrow. (Mixture is stained for illustration purposes only.) Water is gently but rapidly layered over the acrylamide by running down the side of the tube (B). (Right) When glass tubes have been filled, it will be possible to see the top of the acrylamide mixture in the tube (arrow). After 1 hr of polymerization, the bottom cap of the casting tube can be screwed off so that the glass tubes can be removed. (*Never* pull tubes out before removing bottom.) The excess polymerized acrylamide can then be removed with a razor blade and the tubes can be separated and rinsed briefly with distilled water.

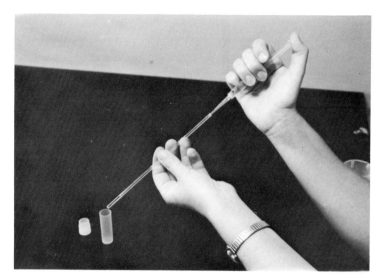

Figure A3.5. Removal of isoelectric focusing gels from glass tubes. Place Eppendorf pipette tip (200-μl size) firmly into top end of IEF and firmly push until gel begins to come out of glass tube. As soon as the gel is about one-half out, release pressure to avoid breaking gel. *Note:* This method is also used for a chamber in which glass tubes are permanently sealed (Fig. A3.7).

2. The tubes can easily be removed by inserting a yellow Eppendorf pipette tip, attached to a 3-ml syringe filled with water, into the top of the tube and gently pushing out the gel (see Fig. A3.5). If samples are difficult to remove, it is possible that you did not clean the tubes adequately with Chromerge. It is not generally advisable to coat the tubes with silicone, as this is time consuming, and the possibility exists that gels may slip out during the focusing, allowing the chambers buffers to mix. This can result in electrophoresis apparatus damage as well as loss of all samples.

F. Procedure for Isogel Equilibration

The IEF gels are equilibrated in buffer for 15 min to remove ampholytes and urea and to recoat the proteins with SDS. In some instances, we have equilibrated the gels for as little as 2–5 min with excellent results. The excess buffer can be easily removed by pouring the gel onto a 50-

Figure A3.6. **Removal of excess equilibration buffer from isoelectric focusing gel.** Drain equilibrated IEF gel on a Nitex nylon screen. The gel is now ready for loading onto the second-dimension gel.

μm Nitex nylon (Tetko, Inc., Elmsford, NY) screen (made by cutting off a 15-ml conical centrifuge tube and by cutting a hole in the screwcap used to hold the screen in place) (see Fig. A3.6). (This buffer can be prepared and stored frozen in small aliquots). Always add 0.5–0.8% β-mercaptoethanol just before use; 1 ml of equilibration buffer per IEF gel is adequate.

II. CASTING MULTIPLE GELS USING SPECIALIZED MULTIPLE ISOELECTRIC FOCUSING GEL APPARATUS

A. Reagents

See Section I.A for the list of reagents.

B. Equipment

Commercial IEF units are used for multiple casting as well as electrofocusing (Health Products, or Electronucleonics) (Figs. A3.7 and A3.8).

Figure A3.7. Casting multiple isoelectric focusing gels, step 1. (A) Multiple casting of IEF using apparatus with permanent glass tubes (model available from Health Products). Before use, the glass tubes are submerged in a 1- or 2-liter beaker of Chromerge. (*Note:* Do not place Plexiglas portions in this solution.) To begin casting gels, place Plexiglas bars (see Fig. A3.8 for closeup) at bottom of glass tubes (arrow).

C. Stock Solutions and Reagents

The stock solutions and reagents are the same as those described in Sections I.A and I.C, except that final acrylamide–ampholyte recipe should be doubled for 40 IEF gel chamber.

D. Procedure for Casting IEF Gels

      1–4. Same procedure as described in Section I.C.
        5. Fill bottom chambers with acrylamide mixture (Fig. A3.9).

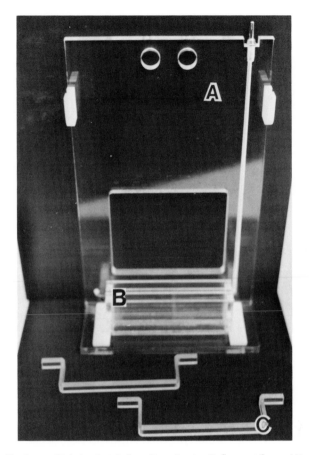

Figure A3.8. Casting multiple isoelectric focusing gels, step 2. Support frame (A), acrylamide wells (B), and Plexiglas bars (C) used for casting multiple IEF gels. (*Note:* The two separate acrylamide wells allows for the simultaneous casting of gels with different ampholyte mixtures.)

6. Overlay acrylamide with water (Fig. A3.10).
7. Gently lower tube apparatus into chamber containing water (Fig. A3.11).
8. Allow gels to polymerize for 1 hr.
9. Remove acrylamide chambers and rinse off gels.
10. Prefocus and load protein samples as described in Section I.C.

E. Procedure for IEF

Same as Section I.E.

Figure A3.9. Casting multiple iso-electric focusing gels, step 3. The glass tube apparatus is placed on the support frame with each row of glass tubes in an acrylamide well. The acrylamide mixtures are then poured into the wells (using a small beaker) immediately after the addition of ammonium persulfate and TEMED.

III. NONEQUILIBRATION ISOELECTRIC FOCUSING (NEPHGE Gel System)

O'Farrell, P. Z., Goodman, H. M., and O'Farrell, P. H., 1977, High-resolution two-dimensional electrophoresis of basic as well as acidic proteins, *Cell* **12**:1133–1142.

A. Reagents

All reagents are the same as for standard equilibration IEF electrophoresis, as described in Section I.A.

Figure A3.10. Casting multiple isoelectric focusing gels, step 4. Water is gently layered over the acrylamide solution in each of the wells.

B. Equipment and Supplies

Equipment and supplies are the same as for standard equilibrium IEF electrophoresis, as described in Section I.B.

C. Stock Solutions

See under equilibration IEF (Sections C.1, C.2, and C.3).

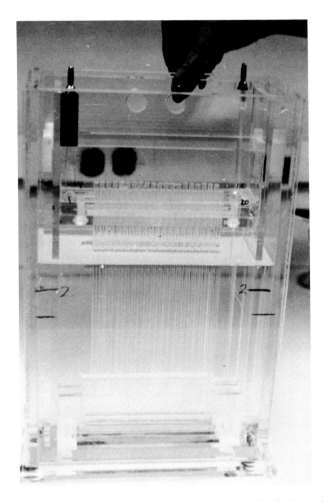

Figure A3.11. Casting multiple isoelectric focusing gels, step 5. Gently lower the tube apparatus into the chamber of water. The level of water in the chamber determines how high the acrylamide will rise in the tubes. This should be determined before starting the gel casting procedure.

D. Procedure

1. Samples for nonequilibrium IEF must be solubilized in the urea solubilization buffer as described in Appendix 2 (Section I.C.2).
2. All gel casting procedures should be carried out as for standard equilibrium IEF, except that the upper and lower buffers are re-

versed. The upper electrode buffer should contain phosphoric acid, and the lower buffer should contain the sodium hydroxide.

3. When attaching the electrodes to the power supply, be sure to attach the upper buffer reservoir to the positive electrode and the lower buffer reservoir to the negative electrode.

4. Finally, the IEF gels should be removed at intervals such as 2000, 4000, 6000, or 8000 V-hr. Total volt-hours will have to be optimized to resolve different proteins of interest because this is a nonequilibrium system.

Preparation of Slab Gels for One- or Two-Dimensional Polyacrylamide Sodium Dodecyl Sulfate Gel Electrophoresis

I. CASTING AND RUNNING INDIVIDUAL ONE-DIMENSIONAL SODIUM DODECYL SULFATE–POLYACRYLAMIDE (NONGRADIENT) GELS

Laemmli, U. K., 1970, Cleavage of structural proteins during the assembly of the head of bacteriophage T4, *Nature (Lond.)* **277**:680–688.

A. Reagents

1. Acrylamide (BioRad, Polysciences, Serva Fine Chemicals, or Sigma)*; the latter two are more inexpensive but will require filtering by Whatman #3 filter paper, then with 0.2-μm Millipore filter. These impure reagents may also have contaminants detectable by silver stain, which can easily be distinguished if two-dimensional PAGE is used but which will interfere if one-dimensional PAGE is used with silver-staining methods.
2. N,N'-Methylene (bis)acrylamide (BioRad)
3. Trizma Base (Sigma)
4. Glycine (Sigma)
5. SDS (BioRad)
6. β-Mercaptoethanol (BioRad)
7. Glycerol (Fisher)

*Manufacturers in parentheses

8. N,N,N',N'-Tetramethylethylenediamine (TEMED) (BioRad)
9. Ammonium persulfate (BioRad)
10. Bromophenol blue (BioRad)
11. Sec-butanol (Fisher)
12. Agarose (BioRad)

B. Equipment and Supplies

1. Glass plates and spacers: The size of plates you use will depend on the type of electrophoresis chamber that will be utilized (e.g., 18 cm × 16 cm plates with 1.5-cm spacers are compatible with BioRad or Hoefer electrophoresis units). The recipes in this Appendix are for these size gels.
2. 50-ml glass beakers
3. 50-ml syringe
4. 200-ml lyophilization flask (optional)
5. Plastic squeeze bottle or aspirator bottle
6. Parafilm
7. Envelope clamps
8. 2-ml glass screwcap tubes
9. Hamilton syringe
10. Electrophoresis chambers can be obtained commercially from Bethesda Research Laboratories (model V16), BioRad (model Protean II), Hoefer (model SE600) or Studier apparatus custommade (Studier, F. W., 1973, Analysis of bacteriophage T7 early RNAs and proteins on slab gels, *J. Mol. Biol.* **79**:237–248) (Fig. A4.1).
11. Electrophoresis power supply: Most power supplies will work for this type of electrophoresis. Some power sources have been designed to be used for both IEF and large- and small-scale PAGE (e.g., Health Products Power Supply) and are therefore more adaptable.

C. Stock Solutions

1. Bis-acrylamide stock
   a. 30% acrylamide, 30 g
   b. 0.8% bis-acrylamide, 0.8 g
   c. Final volume to 100 ml (filter to 0.2 μm for silver staining)

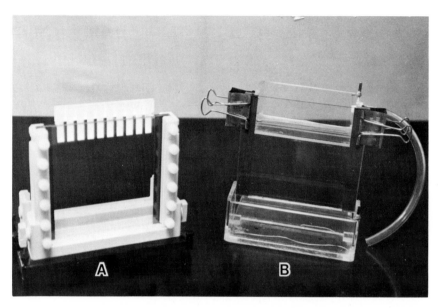

Figure A4.1. Equipment commonly used to cast and run individual slab polyacrylamide gels.
(A) Slab gel-casting apparatus (BioRad). (B) Electrophoresis apparatus described by Studier
(1973).

2. Running gel buffer stock
   a. 1.5 M Trizma Base, 182 g
   b. 0.4% SDS, 4 g
   c. Final volume to 1000 ml, pH 8.8
3. Stacking gel-buffer stock
   a. 0.5 M Trizma base, 30 g
   b. 0.4% SDS, 2 g
   c. Final volume 500 ml, pH 6.8
4. Tank buffer
   a. 0.025 M Trizma base, 15 g
   b. 0.192 M glycine, 72 g
   c. 0.1% SDS, 5 g
   d. Make final volume to 5 liters with distilled H_2O. You may make
      up packets of powder in advance and dilute just before use.
      This volume is compatible for use with the BioRad or Hoeffer
      Electrophoresis tank.

D. Sample Preparation for One-Dimensional SDS–PAGE

    1. Add solubilization buffer to sample. In general, 200 μl buffer can be added to 10–500 μg protein for good results (be sure to optimize so that the SDS : protein ratio is adequate, as described in Chapter 3). If in doubt, it is generally advisable to add extra solubilization buffer.

    2. Heat sample for 5–10 min at 95°C in a boiling waterbath. Be sure to use glass tubes, as plastic tubes will insulate and prevent adequate heating. Also, screwcap tubes are recommended to ensure that solution does not evaporate during heating. It may also be adequate and sometimes advantageous to allow samples to incubate for 2–3 hr at room temperature. However, if proteases are present that are active in the presence of SDS, this may not be the best method.

E. Procedure for Casting Running Gel

    1. Assemble the appropriate clean dry glass plates and spacers (Fig. A4.1).

    2. Combine acrylamide, buffer, and H_2O for the running gel; then degas. The use of a lyophilizer ensures complete degassing.

	7.5%	10%	12.5%	15%
Bis-acrylamide stock	7.1 ml	9.5 ml	11.8 ml	14.2 ml
Gel buffer	7.1 ml	7.1 ml	7.1 ml	7.1 ml
H_2O	14.2 ml	11.8 ml	9.5 ml	7.1 ml
10% Ammonium persulfate	10.5 μl	105 μl	105 μl	105 μl
TEMED	15 μl	15 μl	15 μl	15μl

    3. Add TEMED and mix thoroughly by swirling beaker. Add ammonium persulfate and swirl gently to mix.

    4. Pour the mixture down one edge of the spacer of the gel casting unit using a 10-ml pipette or a syringe and a large (18-gauge) needle. Fill to within 4 cm of the top of the glass plates if the gel is to be used for one-dimensional separation and to within 3 cm of the top if an IEF gel is to be added.

5. Carefully overlay with water saturated sec-butanol and allow to polymerize 45–60 min. (*Note:* This is the most important step for obtaining good resolution in slab polyacrylamide gels.)
6. When the running gel is polymerized, remove sec-butanol (be sure to use correct butanol). Rinse the gel several times with distilled water, and drain well.

F. Procedure for Casting Stacking Gel

1. The same stacking gel is used regardless of the percentage acrylamide in the running gel. The solution for the stacking gel is as follows:
   a. Bis-acrylamide stock, 1.9 ml
   b. Stacking gel buffer stock, 3.1 ml
   c. H_2O, 7.5 ml
   d. 10% ammonium persulfate, 60 μl
   e. TEMED, 10 μl
2. Combine and degas reagents for the stacking gel. Add TEMED and mix by swirling; then add ammonium persulfate. Swirl to mix and layer on top of running gel. Insert a comb to form wells, being sure not to trap bubbles on the bottom of the comb dividers.
3. Prepare tank buffer. Assemble gel in chamber, and remove air bubbles.
4. Underlayer samples in wells with a Hamilton syringe after the gel plate has been attached to the upper buffer reservoir.
5. Electrophoresis can be carried out at 100 mA/per gel (constant amperage) during the day or as low as 10 mA per gel overnight. You can also use constant voltage or constant power.
6. Stain gels using methods described in Appendix 9.

II. CASTING AND RUNNING INDIVIDUAL GRADIENT GELS

A. Reagents

See Section I.A.

B. Equipment and Supplies

1. Glass plates and spacers (same as for Section I.B)
2. Gradient maker (two-cylinder style)

3. Peristaltic pump and tubing
4. Magnetic stir plate
5. Electrophoresis slab gel apparatus (available from BioRad or Hoefer or easily made according to the method of Studier, 1973). This apparatus is illustrated in Figure A4.1 and is now available commercially from Bethesda Research Laboratories.
6. Electrophoresis power supply

C. Stock Solutions

1. Second-dimension bis-acrylamide stock
   a. 0.8% bis-acrylamide, 1.6 g
   b. 30% acrylamide, 60 g
   c. Dissolve in H_2O, 140 ml
   Mix thoroughly and bring to 200 ml with distilled H_2O (or use Continental Water System deionized filtered H_2O). Filter with Whatman #2 filter paper. If silver staining is to be used, it is advisable to then filter to 0.2 μm. This stock can be stored for a few weeks at 4°C.
2. Second-dimension buffer stock
   a. Trizma base, 40 g
   b. Trizma HCl, 20 g
   c. Bring to 300 ml
   d. Adjust pH to 8.5–8.6
3. 10% second-dimension buffer
   a. 3 parts second-dimension buffer stock (see Section II.C.2)
   b. 5 parts H_2O
4. 20% second-dimension buffer
   a. 3 parts second-dimension buffer stock (C2 above)
   b. 1 part glycerol
5. Ammonium persulfate
   a. 10% ammonium persulfate, 10 g
   b. H_2O, 100 ml
   Freeze at -20°C in small aliquots to guarantee the consistency of your polymerization for as long as your stock lasts. (*Note:* Depending on the amount of hydration of your ammonium persulfate, different stock solutions may vary.) You may therefore have to adjust the amount of stock accordingly. Dramatic changes in room temperature will also affect the rate of polymerization so that the amounts of each of these will have to be adjusted. In general, the amounts of initiator (ammonium persulfate) and catalyst (TEMED)

should be optimized carefully to yield approximately 7–10 min working time before the top of the gels polymerize and 15–30 min before gelation occurs at the bottom of the gradient.

6. 10% SDS
   a. SDS, 10 g
   b. H_2O, 100 ml. If silver staining methods are to be used, all reagents should be filtered to 0.2 μm.

D. Procedure

1. Seal and cast individual slab gels. Commercially available slab gel apparatus can be used to cast gels (as illustrated in Fig. A4.2). Gel plates are sealed together between clamps provided with the chamber. Alternatively, glass gel plates can be sealed using large envelope clamps and parafilm, as illustrated in Figure A4.2. To seal plates:
   a. Place spacers between plates and clamp with envelope clamps
   b. Fold parafilm tightly over bottom and clamp

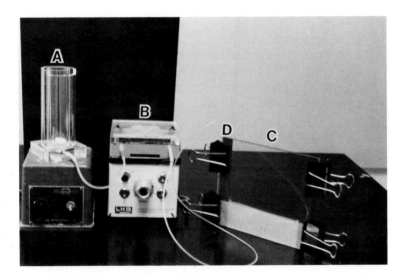

Figure A4.2. Apparatus for casting individual acrylamide gradient-slab gels. (A) Double-chamber conical gradient maker. (B) Peristaltic pump. (C) Glass plates sealed with parafilm at the bottom and by running hot agarose down the sides between the spacers before clamping with envelope clamps. (D) Tubing placed at top edge of glass plates while pouring gradient.

Note: Although it is sometimes easier to use vacuum seal grease to help seal spacers and plates, this usually results in difficulty in cleaning plates, and silver-staining artifacts are increased. We recommend that vacuum seal grease not be used for this reason.

    c. Very hot 1% agarose must be run down both sides of the prepared gel plate. This agarose should be allowed to penetrate along the bottom of the plate and can also be used on the outer edges of the glass plates next to the spacers to ensure a good seal.

2. Pour individual gradient gels.

    a. For gradient gel (approximately 40 ml vol), prepare the following reagents:

	Second-dimension bis-acrylamide stock (ml)	10% second-dimension buffer (ml)	20% second-dimension buffer (ml)	10% SDS (ml)	10% persulfate ammonium (μl)	TEMED (μl)
10% gradient mix	6.3	14.7	0.0	0.2	30	2
20% gradient mix	14.7	0.0	6.3	0.2	15	2

    b. Close gradient chamber.

    c. Place 20% gradient mix in internal chamber and begin mixing with magnetic stir bar. (The stir bar should be large enough to fill the chamber and spin fast enough for vigorous mixing without the formation of excessive numbers of air bubbles in acrylamide.)

    d. Place 10% gradient mix in external chamber.

    e. Start peristaltic pump and open gradient chamber to begin gradient formation.

    f. When gradient is finished, spray surface of gel with water-saturated sec-butanol.

    g. The addition of a stacking gel is not essential if gradient gel electrophoresis is used for a two-dimensional gel. It is useful if you need to form sample wells, as shown in Figure A4.2.

    h. Rinse surface of gel with H_2O and load sample(s) or IEF gel as described above and in Appendix 5.

    i. Carry out electrophoresis at 10–120 mA (constant current). Gels can be run at slow rates at low current overnight if desired.

III. CASTING AND RUNNING MULTIPLE GRADIENT GELS

A. Reagents

See Section I.A.

B. Equipment

1. Multiple casting chamber (see Chapter 11) (available from Health Products or Integrated Separation Systems or Electronucleonics) (Fig. A4.5).
2. Electrophoresis chamber for running multiple gels
3. Magnetic stir plate
4. Power supply (capable of reaching 1.5 A), available from Health Products or Electronucleonics
5. Preformed glass plates with spacers

C. Stock Solutions

1. Second-dimension bis-acrylamide stock
   a. 0.8% bis-acrylamide, 16 g
   b. 30% acrylamide, 600 g
   c. Dissolve in 1400 ml H_2O
   d. Mix thoroughly and bring to 2 liters with distilled H_2O (or use Continental Water System deionized filtered H_2O).
   e. Filter with Whatman #2 filter paper. If silver staining is to be used, it is advised to then filter to 0.2 μm. The apparatus for large scale filtration of acrylamide is shown in Figure A4.3. This stock can be stored for a few weeks at 4°C.
2. Second-dimension buffer stock
   a. Trizma base, 400 g
   b. Trizma HCl, 200 g
   c. Bring to 3 liters
   d. Adjust pH to 8.5–8.6
3. 10% second-dimension buffer
   a. 3 parts second-dimension buffer stock (#2 above)
   b. 5 parts H_2O
4. 20% second-dimension buffer
   a. 3 parts second-dimension buffer stock (#2 above)
   b. 1 part glycerol

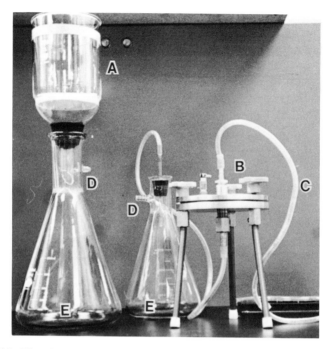

Figure A4.3. Filtration setup for preparation of large quantities of acrylamide for second-dimension electrophoresis. (A) Scintered glass filter with Whatman #3 filter paper for crude filtration. (B) Millipore tissue culture medium filter for filtration of acrylamide to 0.22 μm. (C) Line from flask containing acrylamide. (D) Hookup to vacuum pump. (E) Filtered acrylamide.

5. Ammonium persulfate
   a. 10% ammonium persulfate
   b. H_2O, 100 ml
   Freeze at $-20°C$ in small aliquots. This will guarantee the consistency of your polymerization for as long as your stock lasts. (*Note:* Depending on the amount of hydration of your ammonium persulfate, different stock solutions may vary.) You may therefore have to adjust the amount of stock used accordingly. Dramatic changes in room temperature will also affect the rate of polymerization so that the amounts of each of these will have to be adjusted. In general, the amounts of initiator and catalyst should be optimized carefully to yield approximately 7–10 min working time before the top of the gels polymerize and 30–60 min before gelation occurs at the bottom of the gradient.

6. Sodium dodecyl sulfate
   a. 10% SDS, 10 g
   b. H$_2$O, 100 ml
7. Electrode buffer
   a. 0.025 M Trizma base, 39 g
   b. 0.192 M glycine, 187 g
   c. 0.1% SDS, 13 g
   Individual packets can be made up in advance. Mix in approximately 1 liter H$_2$O on magnetic stirrer until dissolved. Bring up to 13 liters directly in electrophoresis tank for final concentration. This recipe is for the Electronucleonics tank. The Mega-DALT from Health Products holds approximately 22 liters. If you plan to use the silver stain, it is advisable to make this up fresh for each run.

D. Procedure (see Figs. A4.4–A4.12)

1. Prepare glass plates by washing in SDS, followed by rinsing with H$_2$O, then ethanol.
2. Load gel-casting chamber with gel plates and seal.
3. Measure out and mix together acrylamide and buffers. Degas thoroughly. Add 10% SDS. (Add ammonium persulfate and TEMED immediately before pouring gradient.) Recipe for casting 10 gradient gels (Pace Linear Gradient Maker and the DALT or MEGA casting chambers; if larger casting chambers or thickness of glass plates vary, recipe will have to be reduced or increased accordingly):

	Second-dimension bis-acrylamide stock (ml)	10% second-dimension buffer (ml)	20% second-dimension buffer (ml)	10% SDS (ml)	10% persulfate ammonium (ml)	TEMED (μl)
10% gradient mix	135.0	270.0	0.0	4.0	5	130
20% gradient mix	270.0	0.0	135.0	4.0	2	15

4. Pour 10% solution into center of gradient maker and add stir bar, which almost completely fills diameter of chamber. Turn on magnetic stirrer until the surface of the acrylamide starts to funnel downward, taking care that air bubbles do not form.

5. Pour 20% solution into outer well.

6. Open clamps to casting chamber and begin pouring gradient. Immediately open gradient chambers to allow 20% acrylamide solution to mix with 10% acrylamide solution (see Fig. A4.6). Rotate chamber slowly as acrylamide reaches top corner of glass plates.

7. When chamber is almost filled with acrylamide, switch to glycerol–dye line and fill remainder of chamber (see Fig. A4.7).

8. Spray surface of chamber generously with water-saturated sec-butanol (see Fig. A4.8).

9. Allow gels to polymerize and cool for 1 hr.

10. Gels can be used immediately or washed and stored at room temperature until use. For best results, gels should be used soon after casting.

Figure A4.4. Glass plates with permanent glass spacers (S) and rubber hinges (H) for second-dimension gel electrophoresis. Plates should be washed with water followed by ethanol before being used.

Figure A4.5. Preparation of gel-casting chamber. Place glass plates in chamber and hold in place until front cover is screwed in place. Gently screw on all wing nuts and then tighten *evenly* until rubber gasket is tightly sealed.

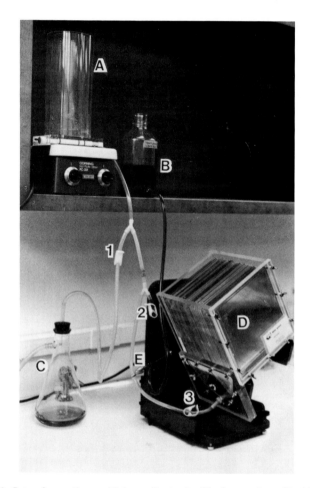

Figure A4.6. Setup for casting multiple-gradient gels. (Equipment from Health Products.) (A) Gradient maker. (B) Water–glycerol (50 : 50) with bromophenol blue. (C) Vacuum flask. (D) Casting chamber filled with glass plates. (E) Screw for tilting chamber. (F) Tubing clamps: 1, 2, and 3.

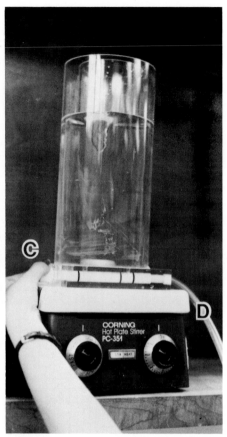

Figure A4.7. Adding acrylamide solutions to gradient chamber. To prepare gradient maker, close valve at bottom of chamber. Add catalysts to both acrylamide solutions after degassing thoroughly. Mix vigorously and pour higher-percentage acrylamide into outside chamber and lower-percentage acrylamide into center chamber.

Figure A4.8. Opening gradient chamber to start flow of acrylamide solutions. Place gradient maker on magnetic stirrer on shelf (approximately 2 ft from bottom of gel plates) and center magnetic stir bar. (*Note:* Stir speed and gradient maker distance should be kept constant for reproducible gradients.) Open chamber valve C and gently press tubing D to remove air bubbles as acrylamide enters tubing.

Figure A4.9. Filling of multiple gel-casting chamber. Fill chamber until acrylamide level (arrows) is at left corner of chamber. Gradually tilt chamber using crank as acrylamide continues to fill chamber until the unit is level.

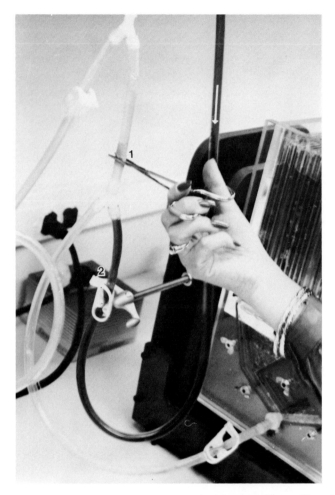

Figure A4.10. Continuation of filling multiple gel-casting chamber. Clamp off acrylamide (1) and open clamp (2) to fill tubing and bottom of chamber with the glycerol solution. This solution should come just up into the chamber at the bottom of the glass plates. (*Note:* This will keep acrylamide from polymerizing in the tubing in the casting chamber.)

Figure A4.11. Cleaning multiple gel-casting chamber tubing. Clamp off tubing next to chamber (1) and backwash tubing with water in gradient maker using vacuum system (see Fig. A4.6,C).

Figure A4.12. Overlay gels with sec-butanol. Spray surface of gels with water-saturated sec-butanol. This will seal the acrylamide from the air so that the gel will polymerize with a smooth surface.

IV. NONGRADIENT SDS–PAGE FOR IMPROVED RESOLUTION OF LOW-MOLECULAR-WEIGHT PROTEINS

Modified from Porzio, M. A., and Pearson, A. M., 1977, Improved resolution of myofibrillar proteins with sodium dodecyl sulfate (SDS) polyacrylamide gel electrophoresis, *Biochem. Biophys. Acta* **490**:27–34.

A. Reagents

    1. Acrylamide (BioRad)
    2. Bis-acrylamide (BioRad)
    3. Trizma-base (Sigma)
    4. Glycine (Sigma)
    5. Glycerol (Fisher Scientific)
    6. SDS (BioRad)
    7. Ammonium persulfate (BioRad)
    8. Bromphenol blue (BioRad)
    9. Deionized and/or double-distilled H_2O

B. Equipment

    1. Slab-gel electrophoresis apparatus (e.g., Studier, 1973, Figs. A4.1 and A4.2); BioRad or Hoeffer apparatus
    2. Electrophoresis power supply
    3. Glass plates and clamps

C. Stock Solutions

    1. 25% acrylamide: 0.25% Bis-acrylamide stock
      a. Acrylamide, 25 g
      b. Bis-acrylamide, 0.25 g
      c. Bring to 100 ml with H_2O
    2. 2 M Tris-glycine buffer (pH 8.8)
      a. 0.5 M Trizma base, 6.05 g
      b. 1.5 M glycine, 11.25 g
      c. Bring to 100 ml with H_2O
    3. 50% glycerol stock
      a. Glycerol, 50 ml
      b. Bring to 100 ml with H_2O

4. SDS–EDTA stock
   a. 2.5% SDS, 2.5 g
   b. 2.5 mM EDTA, 0.095 g
   c. Bring to 100 ml with H_2O
5. 1% TEMED stock
   a. TEMED, 1 ml
   b. Add H_2O, 99 ml
6. 1% ammonium sulfate stock
   a. Ammonium persulfate, 1 g
   b. Bring to 100 ml with H_2O
7. 10% SDS stock
   a. SDS, 10 g
   b. Bring to 100 ml with H_2O
8. 10% ammonium persulfate stock
   a. Ammonium persulfate, 10 g
   b. Bring to 100 ml with H_2O
9. Electrophoresis chamber buffer: Tris/glycine (pH 8.8)
   a. Dilute 10 ml Tris-glycine stock (C2 above) to 1 liter with H_2O
   b. Add 1 g SDS

D. Procedure

1. Prepare samples.
2. Assemble glass plates for casting gel, as shown in Figure A4.1.
3. Prepare mixture for polymerization of 10% acrylamide gel. Recipe for 10% acrylamide gel (adequate for one 30-ml gel, which is for standard 16 × 18 cm plates, leaving room for stacking gel):
   a. Acrylamide stock solution, 12 ml
   b. Tris-glycine buffer, 6 ml
   c. Glycerol stock, 3 ml
   d. SDS–EDTA stock, 1.2 ml
   e. H_2O, 5.4 ml
   f. TEMED stock, 1.2 ml
   g. Ammonium persulfate stock, 1.2 ml
4. Pour acrylamide mixture into glass plates.
5. Overlay acrylamide mixture immediately with layering solution I:
   a. Tris-glycine buffer, 20 ml
   b. TEMED stock, 40 μl
   c. 10% ammonium persulfate stock, 0.4 ml
   d. Bring to 100 ml with H_2O

6. Allow gel to polymerize for 20 min.
7. Remove first overlay solution, rinse, and overlay with layering solution II.
   a. Tris-glycine buffer, 20 ml
   b. Glycerol stock, 5 ml
   c. 10% SDS stock, 0.5 ml
   d. Bring to 50 ml with H_2O
8. If you are preparing a slab acrylamide gel, you will need to pour a gel for forming sample wells. A stacking gel can be used for this purpose. Sample and stacking gel can be prepared as described for the Laemmli procedure (Appendix IF above).
9. Allow gel to polymerize for 1 hr.
10. Initiate electrophoresis at 5–7 mA until samples enter gel. Carry out electrophoresis at approximately 12 mA for 4–6 hr. (*Note:* amperage and voltage may vary according to size and numbers of gels.)

Loading the Isoelectric Focusing Gel and Carrying out Second-Dimension Electrophoresis

I. LOADING THE ISOELECTRIC FOCUSING GEL ONTO THE SECOND-DIMENSION GEL

A. Reagents

1. Agarose (BioRad)*
2. Trizma base (Sigma)
3. Sodium dodecyl sulfate (Sigma)
4. H_2O (deionized or distilled)
5. Glycine (Sigma)

B. Equipment and Supplies

1. Gel-loading platform (Electronucleonics or Health Products)
2. Pasteur pipette and bulb
3. Spatula

C. Stock Solution: Overlay Agarose Recipe

1. 0.025 M Trizma base, 3 g
2. 0.192 M glycine, 14.4 g
3. 0.1% SDS, 1 g
4. 0.5% agarose, 5 g

*Manufacturers in parentheses.

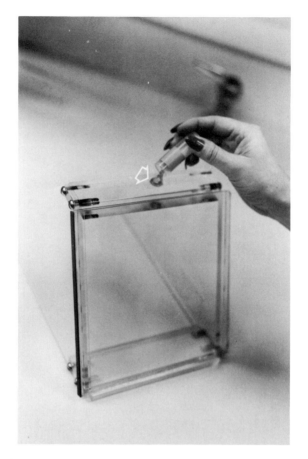

Figure A5.1. Loading the isoelectric focusing gel, step 1. The gel (arrow), which has been equilibrated with the second-dimension equilibration buffer, is poured from its plastic vial onto the loading platform. The second-dimension gel is placed onto the platform.

Fill to 1 liter with H_2O and heat in microwave or in boiling water bath until agarose is *completely* in solution. Freeze in 50-ml conical centrifuge tubes. Agarose can be liquified by heating in a boiling water bath for everyday use.

D. Procedure

1. Place slab gel plate on loading platform (Fig. A5.1).
2. Lay IEF gel on platform and gently straighten out (Figs. A5.1–A5.3).

3. One end of the gel is dropped between the glass plate onto the surface of acrylamide of the gradient gel. The rest of the isoelectric focusing (IEF) gel should slide into place along the surface of the sodium dodecyl sulfate–polyacrylamide gel electrophoresis (SDS–PAGE) gel. If the IEF gel does *not* easily slip between the glass plates onto the surface of the acrylamide, you need to readjust the thickness of either your IEF or slab gel. *Be sure that no air bubbles are trapped between IEF gel and surface of acrylamide slab gel.* The quality of the second-dimensional gel will depend in part on the IEF gel interface with the gradient gel.
4. Seal the isogel with a small amount of overlay agarose. Again, make sure that there are no air bubbles.

II. CARRYING OUT SECOND-DIMENSION ELECTROPHORESIS

A. Electrophoresis of Individual Gels

If standard electrophoresis chambers are used (e.g., Studier, BioRad, or Hoeffer electrophoresis apparatus), electrophoresis is carried out by

Figure A5.2. Loading the isoelectric focusing gel, step 2. The gel (arrows) is spread out across the loading platform using a small flat spatula. Excess fluid is removed from gel.

Figure A5.3. Loading the isoelectric focusing gel, step 3. The gel is moved from the platform to the back edge of the gel plate. (1) One end of the IEF gel is moved onto the surface of the polyacrylamide. If the first-dimension IEF gel is the correct diameter, it will easily slip between the gel plates and come into close contact with the acrylamide. Care should be taken to ensure that no air bubbles are trapped between the IEF gel and the acrylamide slab gel. (2) A small piece of agarose containing molecular-weight standards is then placed adjacent to the isoelectric focusing gel at one end of the slab gel. (3) Use a glass Pasteur pipette to cover surface of gels with the hot agarose overlay solution. Allow this agarose to cool before loading gels into electrophoresis chamber.

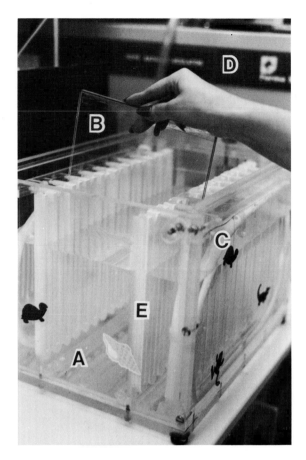

Figure A5.4. Loading slab gels into electrophoresis. (Chamber from Electronucleonics.) Gels are placed between rubber slits. Chamber is filled with electrophoresis buffer. (A) Electrophoresis chamber. (B) Slab acrylamide gel. (C) Cooling coil. (D) Cooling bath for recirculating buffer cooling system (Forma Scientific). (E) Rubber slits.

placing electrode buffer in upper and lower chambers. The slab gels are then placed into these chambers, with care being taken to avoid air bubbles being trapped at the bottom of the slab acrylamide gel. This can be done by tilting the gel as it is lowered into the chamber and by tilting the chamber so that the buffer will move across the bottom of the gel to remove trapped air bubbles.

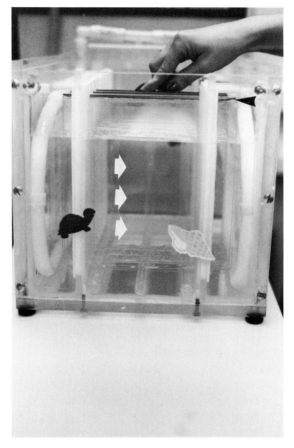

Figure A5.5. Second-dimension gel electrophoresis of multiple gels. Gels are placed in chamber to fill all slots. If too few gels are available, thick glass or plexiglass plates can be used to block empty slots. Water is then added to fill chamber to top of glass plates (black arrow). Electrophoresis is carried out (direction of arrows) from side of gel containing IEF gel (cathode side) until bromophenol blue dye marker is completely off the gel (anode side).

B. Electrophoresis of Multiple Gels

Electrophoresis chambers for running 10–20 second-dimension gels simultaneously are available from Health Products, Integrated Separation Sciences, and Electronucleonics. Buffer solutions are prepared and added to tanks. (*Note:* For convenience, a small volume of buffer may be made from buffer packets prepared in advance and the remainder of the volume of water can be added directly to the electrophoresis tanks.)

The slab gels are then placed on their sides and are slipped between the rubber spacers (Fig. A5.4 and A5.5). Electrophoresis is carried out for 3–15 hr, depending on cooling system, at 70–400 V.

Appendix 6

Methods for Studying Protein Phosphorylation

I. 8-AZIDO cAMP PHOTOAFFINITY LABELING FOR ANALYSIS OF cAMP-BINDING PROTEINS

Tash, J. S., Guerriero, V., and Means, A. R., 1986, cAMP- and calcium-dependent protein kinases and phosphoprotein identification, in: *Laboratory Methods Manual for Hormone Action and Molecular Endocrinology* (W. T. Schrader and B. W. O'Malley, eds.), pp. 1–50, Houston Biological Assocation, Inc., Texas.

A. Reagents

1. 8-Azido [^{32}P]-cAMP
2. Potassium phosphate (K$_2$HP0$_4$**code**H$_2$0)
3. Magnesium chloride (MgCl$_2$**code**6H$_2$0)
4. cAMP
5. Glycerol
6. Isoelectric focusing (IEF) or sodium dodecyl sulfate–polacrylamide gel electrophoresis (SDS–PAGE) solubilization buffer (see Appendix 2)

B. Stock Solution: Potassium Phosphate Buffer (pH 6.5)

1. 50 mM potassium phosphate, 1.14 g
2. 1 mM MgCl$_2$**code**6H$_2$0, 0.02 g
3. Bring to 100 ml and pH 6.5

C. Procedure

1. 8-azido [^{32}P]-cAMP (5 μCi) is added to 50 μl potassium phosphate buffer and evaporated in the dark with N$_2$ to the original volume to remove alcohol.
2. Samples to be analyzed are mixed with buffer containing azido [^{32}P]-cAMP.
3. Control samples should include
   a. excess unlabeled cAMP (10 μM)
   b. 8-azido [^{32}P]-cAMP (not to be irradiated)
4. Incubate tubes for 60 min in the dark at 4°C or -20°C in the presence of 20% glycerol (to reduce proteolysis).
5. Irradiate tubes with short-wave UV using a mineral lamp or light gel box at distance of 5 cm for 12–15 min.
6. Solubilize samples in buffer (Laemmli or IEF buffer) for one- or two-dimensional

II. PHOSPHORYLATION OF CELL CYTOSOL PROTEINS WITH [v-^{32}P]-ATP TO IDENTIFY SUBSTRATES FOR cAMP-DEPENDENT PHOSPHORYLATION

Richards, J. S., Sehgal, N., and Tash, J. S., 1983, Changes in content and cAMP-dependent phosphorylation of specific proteins in granulosa cells of preantral and preovulatory ovarian follicles and in corpora lutea, *J. Biol. Chem.* **258**(8):5227–5232).

A. Reagents

1. [v-^{32}P]-ATP*
2. Magnesium chloride (MgCl$_2$**code**6H$_2$0)
3. Potassium phosphate (K$_2$HPO$_4$**code**3H$_2$0)
4. Tissue cytosol
5. Dry ice
6. cAMP (Sigma)*
7. Catalytic subunit of type II cAMP-dependent protein kinase

This can also be prepared according to the method of Reimann *et al.*, 1971.†

*Manufacturers in parentheses.
†Reimann, E. M., Brostram, C. O., Corbin, J. D., King, C. A., and Krebs, E. G., 1971, Separation of regulatory and catalytic subunits of the cyclic 3', 5'–adenosine monophosphate dependent protein kinase(s) of rabbit skeletal muscle, *Biochem. Biophys. Res. Commun.* **42**:187–194.

B. Equipment and Supplies

   1. X-ray film for autoradiography
   2. Tissue homogenizer
   3. Two-dimensional PAGE equipment

C. Stock Solution: Potassium Phosphate Buffer (pH 7.4)

   1. 10 mM potassium phosphate, 0.228 g
   2. 10 mM $MgCl_2$, 0.203 g
   3. Bring to 100 ml with H_2O and pH to 7.4

D. Procedure

   1. Prepare tissue by homogenizing in 0.5–1 ml potassium phosphate buffer (pH 7.4).
   2. Centrifuge sample at 30,000 × g for 30 min at 4°C. Take aliquot for protein determination; freeze sample immediately on dry ice and store at −70°C until assay is to be carried out.
   3. Before assay, dilute cytosol to 100 μg/10 μl and add to 40 μl potassium phosphate buffer containing 20 μCi 10 μM [v-^{32}P]-ATP.
   4. Controls and experimental variables for assay should include the following:
      a. + cAMP (2 μM)
      b. − cAMP (2 μM)
      c. + catalytic subunit of cAMP-dependent protein kinase
      d. − catalytic subunit of cAMP-dependent protein kinase
      e. combination of a with c and d or b with c or d
   5. Incubate tubes for 7–10 min at 30°C.
   6. Stop reaction by adding protein solubilization buffer (Appendixes 2 and 4) for one- or two-dimensional PAGE as described above.
   7. Following electrophoresis, polyacrylamide gels should be silver stained, dried, and exposed to x-ray film to detect phosphorylated proteins. For details on experimental design and interpretation of results, see Chapter 5.

III. *IN VIVO* LABELING OF PHOSPHOPROTEINS OF INTACT CELLS OR TISSUE USING [³²P]ORTHOPHOSPHATE

O'Connor, C. M., Blazer, D. R., Jr., and Lazarides, E., 1979, Phosphorylation of subunit proteins of intermediate filaments from chicken muscle and nonmuscle cells, *Proc. Natl. Acad. Sci. USA* **76:**819–823.

A. Reagents

    1. [³²P]phosphoric acid ($^{32}PO_4{}^{3-}$) (New England Nuclear, carrier free)
    2. Phosphate-free minimal medium (GIBCO)
    3. Solubilization buffer for IEF SDS–PAGE (Appendix 2)

B. Equipment and Supplies

    1. Safety shields for radiation
    2. Ground-glass homogenizer
    3. Tubes for sample centrifugation (Appendix 2)

C. Stock Solutions

    1. Neutralize [³²P]phosphoric acid (for direct injection into animal tissue)
    2. [³²P]phosphoric acid for excised tissue or cell cultures. Dilute ($^{32}PO_4{}^{3-}$) with phosphate-free minimal essential medium to give final concentration of 100–200 μCi/ml (tissue) or 60–100 μCi/ml (cell culture).

D. Procedure

    1. Labeling tissue *in vivo* (e.g., chick thigh muscle)
       a. Inject directly with 0.5 mCi neutralized $^{32}PO_4{}^{3-}$.
       b. Following 24 hr, excise, mince, and homogenize tissue in 3 vol solubilization buffer with ground-glass homogenizer.
       c. Centrifuge to remove insoluble material and analyze proteins by one- or two-dimensional electrophoresis.
    2. Labeling tissue *in vitro*
       a. Excise tissue and cut into small pieces (~1 mm).
       b. Equilibrate tissue with phosphate-free minimal essential media for 30–60 min.

    c. Incubate tissue in 0.5 ml phosphate-free minimal essential medium containing 100–200 μCi/ml $^{32}PO_4^{3-}$ for 4–5 hr.

    d. Remove tissue and rinse in phosphate-free minimal essential medium.

    e. Homogenize and process tissue as described above.

3. Labeling cells cultured *in vitro*

    a. Cells should be plated and grown according to experimental conditions that are under investigation (e.g., primary myogenic cell cultures used by O'Connor *et al.*, 1979, were plated at an initial density of 0.6–0.8 × 10^6 per 35-mm plates).

    b. Add 1 ml phosphate-free MEM (−serum) for 1 hr.

    c. Remove media.

    d. Add 0.8 ml phosphate-free minimal essential medium without serum containing 60–100 μCi/ml $^{32}PO_4^{3-}$ to a 35-mm plate and incubate for 4–6 hr at 37°C (or 1 mCi/ml for 1–2 hr or less) ± simulators of phosphorylation).

    e. Wash cells with phosphate-free MEM and solubilize cell monolayer by adding solubilization buffer directly to plate.

    f. Analyze phosphoproteins by electrophoresis.

4. Stain gels with Coomassie blue or silver stain and dry gels (Appendix 8) for autoradiography.

5. Carry out autoradiography using Kodak NS-2T no-screen film at room temperature or Kodak X-Omat XR5 film with duPont Cronex intensifier screen at −70°C.

6. Develop no-screen film with Kodak D-19 developer and X-Omat film with Kodak x-ray developer or utilize automatic processor if available.

IV. ^{125}I-LABELING OF CELL-SURFACE PROTEINS OF CELLS IN CULTURE

Keil Dlouha, V., and Darmon, M., 1983, Changes in pattern and accessibility for ^{125}I-labeling of cell surface proteins after mesenchymal differentiation of embryonal carcinoma cells, *Biochim. Biophys. Acta* **734**:249–259.

A. Reagents

1. Hank's medium (GIBCO)
2. Glucose (GIBCO), tissue culture grade
3. EDTA (GIBCO), tissue culture grade

    4. Lactoperoxidase, 50–70 U/mg (Sigma)

    5. Carrier-free ^{125}I (sodium salt) (Amersham or New England Nuclear)

    6. Glucose oxidase, 250–300 U/mg (Sigma)

    7. Potassium iodide (KI) (Sigma)

    8. Sodium chloride (NaCl) (Sigma)

    9. Sodium phosphate (Na_2HPO_4) (Sigma)

   10. Potassium phosphate (KH_2PO_4 (Sigma)

   11. Reagents for two-dimensional PAGE and autoradiography

B. Equipment and Supplies

    1. Tissue culture facilities
    2. Radioactive safety hood
    3. Radioactive disposal supplies
    4. Equipment for protein analysis by two-dimensional PAGE and autoradiography

C. Stock Solutions

    1. Cell culture buffer A
      a. 0.2 M NaCl, 116 g
      b. 3.2 mM Na_2HPO_4, 0.454 g
      c. 2.6 mM KCl, 0.194 g
      d. 1 mM EDTA, 0.380 g
      e. Bring to 1 liter with distilled tissue culture water and pH to 7.4
    2. Hank's medium with glucose
      a. Hank's medium, 100 ml
      b. 5.5 mM glucose, 100 mg
    3. Hank's medium with potassium iodide
      a. Hank's medium, 100 ml
      b. 150 mM KI, 2.5 g

D. Procedure

    1. Grow cells in culture to the state dictated by the experiment (e.g., exponential growth versus confluency). Control cell populations may have to be dissociated prior to surface labeling. This can be achieved in some cases by removing culture medium and adding

cell culture buffer A (C1 above) and incubating cells for 7 min at 22°C. Following this incubation, the buffer is removed and cells are washed and detached with Hank's solution containing glucose (see Section IV.C.2).

2. Cells attached to culture plates or dissociated cells should be incubated with 5 ml Hank's solution containing glucose (see Section IV.C.2); the precise volume will depend on the size of the culture plate.

3. 50 μg lactoperoxidase (67 U/mg); 1 mCi of carrier-free ^{125}I and 10 μg glucose oxidase (2 should be added and iodination reaction performed at 25°–27° for 20 min).

4. Attached cells should be detached with EDTA buffer (see Section IV.C.1).

5. All cell samples are then washed with 15 ml Hank's KI solution (see Section IV.C.3).

6. Cells should be pelleted and solubilized with buffer for IEF (Appendix 2).

V. METABOLIC LABELING OF PROTEINS IN SHORT-TERM ORGAN OR CELL CULTURE

A. Reagents

1. L-[^{35}S]methionine, >800 Ci/mmole, translation grade (New England Nuclear)

2. ^{14}C-labeled amino acids—16 amino acids (Amersham)

3. Minimum Essential Medium: Eagle (modified) with Earle's salts without glutamine and methionine (Flow Laboratories) or Dulbecco's modified Eagle's medium (DMEM). Different experimental systems may require other specialized media.

4. L-Glutamine: 5% CO_2/95% O_2 gas mixture

5. Reagents for electrophoretic analysis of proteins (one- or two-dimensional PAGE)

B. Equipment and Supplies

1. Shaking water bath (37°C) (for organ culture)

2. Sterile Erlenmeyer flasks (15–25 ml)

3. Incubator for cell culture

C. Procedure for Labeling Proteins in Organ Culture

1. Heat water bath and MEM media to 37°C.
2. Add L-glutamine (435 mg/liter) and L-[^{35}S]methionine (1 mCi/100 ml) to the media and then filter through a 0.22-μm filter to maintain sterility of media until the addition of the tissue or cells.
3. Add 15 ml prepared media into each sterile Erlenmeyer needed. This amount of media containing 150–500 μCi L-[^{35}S]methionine is generally adequate for a culture of approximately 0.5 g tissue over a period of 12 hr.
4. Place the flasks in the shaking water bath.
5. Fresh tissue is then rapidly minced with fine dissecting scissors. Approximately 1-mm^3 pieces of tissue can be used for metabolic labeling.
6. The tissue pieces are added to the media, and the flasks are filled with a 5% Co_2/95% O_2 mixture, using a rubber hose and sterile pasteur pipette. The flasks are then stoppered and incubated with shaking.
7. It is necessary to periodically ($\sim$30–60 min) replenish the 5% CO_2/95% O_2 mixture in the flasks.
8. Remove samples of the tissue at selected time points and rapidly freeze the aliquots to -70°C for future analysis by immunoprecipitation or electrophoresis.
9. Aliquots of the medium should also be frozen when soluble or secreted proteins are being investigated.
10. Control tissue that is incubated in the absence of [^{35}S]methionine should also be collected immediately upon addition of the tissue pieces to the medium in step 5. These samples should be analyzed by electrophoresis using the silver-staining method to compare with tissue collected following incubation *in vitro*.

D. Procedure for Labeling Proteins in Cell Culture

Modified from Bravo, R., 1984, Two-dimensional gel electrophoresis: A guide for the beginner, in *Two-Dimensional Gel Electrophoresis of Proteins: Methods and Applications* (J. E. Celis and R. Bravo, eds.), pp. 4–34, Academic Press, New York; and Bravo, R., Fey, S. J., and Small, J. V., 1981, Coextistence of three major isoactins in a single Sarcoma 180 cell, *Cell* **25**:195–202.

1. Grow cells in culture. The number in the cell cycle will depend on the experimental question being asked (see discussion in Chapter 6).

2. Prepare media and add [³⁵S]methionine (200–1000 μCi) or amino acids (50 μCi). The media may be supplemented with 10% dialyzed fetal calf serum and 1 ng/liter of cold methionine if necessary for cell survival during longer labeling periods. [*Note:* if short-term (up to 3 hr) labeling is used, DMEM without methionine is used because it increases significantly the radioactivity incorporated into proteins. If long-term labeling is used, this may not be acceptable for your cells.]

3. The amount and volume of label to be added to cell cultures will depend on factors such as the number of cells and the specific activity of precursors (see discussion in Chapter 6). These conditions will have to be optimized for your experimental system.

4. Following the label procedure, cells should be washed to remove excess label and media and can be solubilized directly with electrophoresis solubilization buffer and processed for one- or two-dimensional PAGE.

Appendix 7

Fluorography

I. EN³HANCE METHOD

A. Reagents

1. EN³HANCE autoradiography enhancer (New England Nuclear)*
   (The advantage of this reagent is that it is convenient to use and
   is not as dangerous to handle as other reagents used for fluorography.)

B. Equipment and Supplies

1. X-ray film
2. Glass pan
3. Gel dryer (BioRad or equivalent)
4. X-ray cassettes

C. Stock Solution

EN³HANCE is used directly from the bottle. No dilution is necessary.

D. Procedure

1. After conventional electrophoresis with or without Coomassie
   staining, simply soak the polyacrylamide gel in enough EN³HANCE

---

*Manufacturers in parentheses.

to cover the gel adequately (1 hr at room temperature with optimal shaking is generally adequate).

2. Pour off the EN^{3}HANCE solution and cover the gel with distilled water. Shake for 1 hr. The gel will turn white as the EN^{3}HANCE precipitates in the gel. (This has to be expected!)
3. Dry the gel directly using the apparatus described in Appendix 8.
4. Expose the gel using x-ray film (XAR-5).

II. DIMETHYL SULFOXIDE METHOD

Tollaksen, S. L., Anderson, N. L., and Anderson, N. G., 1984, Operation of the ISO-DALT system, 7th ed., ANL-BIM-84-1. Argonne National Laboratory, Argonne, Illinois.

A. Reagents

1. Dimethyl sulfoxide (DMSO) (New England Nuclear)
2. 2,5-diphenyloxazole (PPO) (New England Nuclear)
3. Glycerol (Fisher) (*Note:* These reagents should be handled with care and disposed of appropriately as regulated by your radiation and biosafety officer. Do not pour these reagents down the sink.)

B. Equipment and Supplies

1. Shaker platform
2. Glass pans
3. X-ray film (XAR-2) (may be preflashed as described in Chapter 6).

C. Stock Solutions

1. DMSO-PPO stock
   a. 13% 2,5'-diphenyloxazane (PPO), 39 ml
   b. Dimethylsulfoxide, 300 ml
   c. DMSO I (DMSO II that has been used three times)
   d. DMSO II (DMSO solution that is used fresh or three times for second incubation of gels)
2. Glycerol solution (2%)
   a. Glycerol, 20 ml
   b. Bring to 1 liter with distilled H$_2$O

D. Procedure

1. Add gels to pan containing dimethylsulfoxide (DMSO I), 250 ml/gel. Depending on the size of pan, up to five gels can be treated at one time. Incubate the gels in this solution for 30 min while shaking at a speed that thoroughly circulates the solution around all gels. (*Note:* This solution can be used up to three times before disposal.)
2. Add gels to a second solution of dimethylsulfoxide (DMSO II) and incubate as above while shaking for 30 min. (This solution can be used as DMSO I after third use).
3. Add gels to DMSO-PPO solution (see Section II.C). Shake thoroughly for 3 hr. (This solution can be used up to 8 times).
4. Place gels in distilled water (250 ml/gel) and shake for 15 min.
5. Change water and shake for an additional 15 min.
6. Pour off water and add 2% glycerol solution (250 ml/gel). Incubate the gels in this solution by shaking thoroughly for 15 min.
7. Change glycerol solution and incubate gels for an additional 15 minutes.
8. Dry gels as described in Appendix 8.
9. Prepare gels for radioautography. XAR-2 film (may be preflashed to linearize the photographic response).
10. Place gels on film into black plastic bags, seal, and store at −80°C for exposure.
11. Develop X-ray film using X-OMAT processor or equivalent process.

Appendix 8

Drying Gels for Autoradiography

A. Equipment

For optimal drying, the slab dryer should be connected to a large mechanical vacuum pump protected with a cold trap and a vapor trap installed in that order between the dryer and the pump. Whatever vacuum source is used, a minimum pressure of 29 inches (733 mm Hg) is required.

1. Gel Dryer Variable Temperature 60°–80°C (BioRad model 483, 35 × 45 cm drying area)*; Ephartec model 433-0100 (Haake) (same size drying area)
2. Dry Ice Cold Trap (Labconco 75350) (Virtis)
3. Vapor Trap with Gas Drying Jar (Fisher 09-204)
4. Drierite Absorbent (American Scientific Products, D8160)
5. Pump (Precision Model D-150; Welch Model 1402; Fisher Maxima model D8A or equivalent)

B. Reagents and Supplies

1. Filter paper (3M chromatography paper; Fisher)
2. Mylar sheet (provided with gel dryer)
3. Polyethylene sheets (provided with gel dryer)
4. Acetic acid (Fisher)
5. Glycerol (Fisher)
6. Polyacrylamide gel to be dried
7. Deionized or distilled H_2O

*Manufacturers in parentheses.

C. Stock Solution: Stock Solution A

   1. 1% glycerol, 10 ml
   2. 10% acetic acid, 100 ml
   3. Bring to 1 liter with H_2O.

D. Procedure

   1. Be sure to photograph gels (either black and white or color) before drying because the gels will lose resolution and will be very dark (if silver stained) after drying.

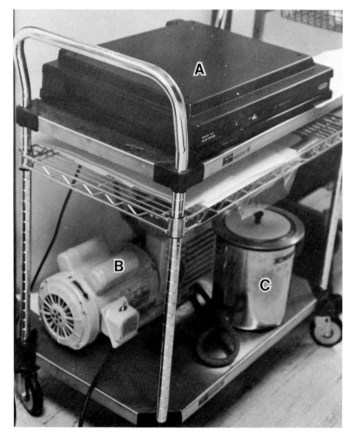

Figure A8.1. Recommended setup for drying polyacrylamide gels. (A) Gel dryer. (B) Vacuum pump. (C) Dry ice cold trap.

2. Preparation of silver-stained gels for drying: These gels must have the sodium carbonate (the last step of the color-based silver stain procedure) exchanged out for 1% glycerol–10% acetic acid by soaking 8–10 hr or overnight. No change is necessary if the soaking solution is five times as much as the volume of the gel.
3. Preparation of Coomassie Blue stained gels for drying: If the polyacrylamide gel has been Coomassie stained and destained, simply rinse with H_2O for a few minutes. (If cracking occurs, you may need to soak the gel in the glycerol–acetic acid solution or adjust the temperature on your gel dryer.)

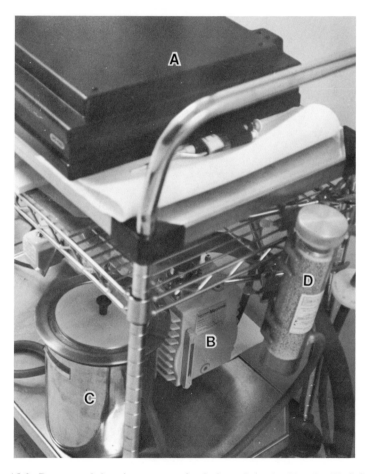

Figure A8.2. Recommended equipment setup for drying polyacrylamide gels. (A) Gel dryer. (B) Vacuum pump. (C) Dry ice cold trap. (D) Vapor trap with Drierite absorbent.

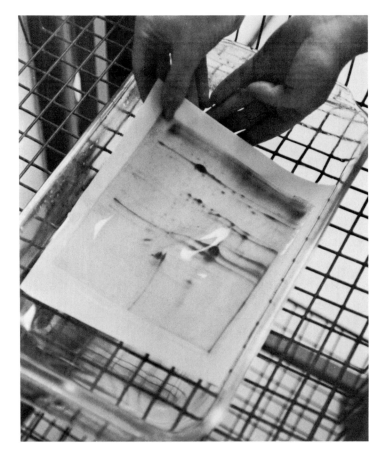

Figure A8.3. Preparation of polyacrylamide gel for gel drying. Filter paper is submerged in water and gel is floated onto paper and air bubbles are removed.

4. Fixed, unstained gels. If the gel has been fixed in 50% methanol : 10% acetic acid but not Coomassie stained, you will probably want to swell the gel back to its original size by soaking in destain (20% methanol : 10% acetic acid) until the gel is back to its original size. Then, carry out steps 1–6.

5. Set up the gel-drying system as outlined in Figures A8.1 and A8.2.

6. Place the gel on a wet filter paper cut slightly larger than the gel in a glass pan with ½ inch of water. Position this gel and filter

Figure A8.4. Drying polyacrylamide gel. (A) Gel is placed on filter paper which covers gel drying screen. (A polyethylene sheet may be used to cover gel here). (B) The Mylar sheet is placed on top of the gel.

paper stack on another wet filter paper already on the gel dryer (Fig. A8.3).

7. Cover the gel with the Mylar sheet or porous polyethylene sheet and remove air bubbles. Do not use plastic wrap on 10–20% gels 1.5 mm thick because it is not porous and the gel will be more susceptible to cracking (Fig. A8.4).

Figure A8.5. Drying the polyacrylamide gel. The gel is covered with the Mylar sheet. The lid is lowered and the vacuum is turned on for 20 min before the heater is turned on to 60°C. After 2 hr, the heater is turned off and the gel is cooled for 15 min with vacuum before it is ready to remove.

8. Begin drying by allowing the vacuum system to run for 20 min before turning on the heat.
9. Apply heat, 60°C for 2 hr.
10. Turn off heat and allow vacuum to run an additional 15 min. Remove dried gel for autoradiography.

Appendix 9

Methods for Staining Proteins in Polyacrylamide Gels

I. COOMASSIE BLUE STAINING METHOD

Modified from Fairbanks, G., Steck, T. L., and Wallach, D. F. H., 1971, Coomassie blue R250 used in isopropanol-acetic acid, *Biochemistry* **10**(13):2606–2618.

A. Reagents

1. Coomassie Brilliant Blue R-250 (BioRad)*
2. Ethanol
3. Acetic acid

B. Equipment and Supplies

1. Glass pyrex dishes
2. Shaking platform (Health Products, South Haven, MI recommended)
3. Filtering apparatus
   a. Funnel for Whatman #2 paper
   b. Nalgene Large Sterilization Filter Unit, Type LS (500 ml; 0.2 μm)
4. Kimwipes

*Manufacturers in parentheses.

C. Stock Solutions

1. Coomassie Brilliant Blue R-250 (0.2%) stock stain solutions
   a. 50% ethanol, 500 ml
   b. 45% H_2O, 450 ml
   c. 5% acetic acid, 50 ml
   d. 0.2% Coomassie Blue, 2 g
   Filter using Whatman #2 filter paper. Filter to 0.2 μm with Nalgene filter if you plan to silver stain gels after you Coomassie Blue stain your gel.
2. Destaining solution
   a. 20% ethanol, 200 ml
   b. 75% H_2O, 750 ml
   c. 5% acetic acid, 50 ml

D. Procedure

1. Add gels to staining solution. Glass pyrex dishes are recommended and should definitely be used if gels are to be silver stained following Coomassie Blue staining. For optimal staining and destaining, samples should be placed on a shaker platform set at a speed to optimally circulate the solution.
2. Destain gels by shaking in changes of destain solution. (*Note:* You can greatly speed the destaining process if you place a handful of Kimwipes on top of the solution. This is more efficient than most electrodestaining procedures or commercially available destaining reagents.)
3. Gels should be photographed using a yellow or red filter for black and white film. Kodak Pan-X film generally gives good results.

II. MONOCHROMATIC SILVER STAIN

Wray, W., Boulikas, T., Wray, V. P., and Hancock, R., 1981, Silver staining of proteins on polyacrylamide gels, *Anal. Biochem.* **118:**197–203.

A. Reagents

1. Silver nitrate ($AgNO_3$) (Sigma)
2. Sodium hydroxide (NaOH) (Sigma)

3. Citric acid anhydrous powder ($H_3C_6H_5O_7$) (Sigma)
4. Formaldehyde, USP 37% (Fisher)

B. Equipment and Supplies

1. Glass pyrex plates ($12 \times 17 \times 3$ inches)
2. Shaking platform (heavy-duty shaker recommended and is essential if large numbers of gels are to be stained at one time: Health Products)

C. Stock Solutions

1. Solution A
   a. Silver nitrate, 0.8 g
   b. Bring to 4 ml with distilled H_2O
2. Solution B
   a. 21 ml 0.36% sodium hydroxide
   b. 1.4 ml 14.8 M ammonium hydroxide (For best results, make fresh to ensure that no evaporation of reagent occurs.)
3. Solution C
   a. Add solution A dropwise to solution B with constant stirring or vortexing
   b. Make to 100 ml with water
   c. Solution C must be used within 5 min
4. Solution D
   a. 1% citric acid, 2.5 ml
   b. 37% formaldehyde, 0.25 ml, made to 500 ml with water (Solution D must be very fresh.)

D. Procedure

1. Soak the gel in 50% reagent-grade methanol at least one hour.
2. Stain gel in solution C for 15 min with constant gentle agitation.
3. Wash gel in deionized water with gentle agitation 5 min.
4. Develop silver stain by soaking gel in solution D until bands appear. Usually, bands appear in less than 10 min and seldom take longer than 15 min.
5. Wash gel with water and place gel in 50% methanol to stop stain development. The basic staining procedure is finished at this point.

6. Some background can be removed by incubating the stained gel in photographic orbit solution.

III. COLOR-BASED SILVER STAIN

Modified from Sammons, D. W., Adams, L. D., and Nishizawa, E. E., 1981, Ultrasensitive silver-based color staining of polypeptides in polyacrylamide gels, *Electrophoresis* 2(3):135–141.

This stain can also be obtained commercially as Gelcode from Health Products or Pierce Chemical.

A. Reagents

1. Silver nitrate ($AgNO_3$) (Sigma)
2. Sodium hydroxide (NaOH) (Sigma)
3. Sodium bicarbonate (Na_2HCO_3) (Sigma)
4. H_2O (distilled, deionized, or double deionized)
5. Formaldehyde, USP 37% (Fisher)

B. Equipment

1. Pyrex glass dishes (12 × 17 × 3 inches) for 8 × 8 inches polyacrylamide gels: Plastic dishes will adsorb many impurities and should never be used for silver staining or immunoblotting.
2. Variable-speed shaking platform (see Fig. A9.1): Shaker from Health Products is recommended because this is a heavy-duty shaker with an extended platform that will hold 200 lb. This shaker is designed for large-scale staining and will not continuously break down routinely (as will most shaker models) if multiple gels are stained routinely.
3. Glass storage bottles: Plastics may leach organics, which will interfere with silver staining, it is recommended that reagents be stored in glass bottles.

C. Stock Solutions

1. Gel fixative—prepare fresh just as needed
   a. 5% acetic acid, 50 ml
   b. 50% ethanol, 500 ml

Figure A9.1. Heavy-duty variable-speed shaker and platform for staining multiple gels. (Available from Health Products.) Multiple pans of gels can easily be stained simultaneously.

    c. 45% H_2O, 450 ml
2. Silver stock
    a. Silver nitrate, 1.9 g
    b. H_2O, 1000 ml
3. Reducing solution
    a. NaOH, 30 g
    b. H_2O, 1000 ml
    c. *Immediately* before using, add 7.5 ml formaldehyde. (*Note:* the age, batch, and source of formaldehyde is critical and should be carefully monitored.)
4. Color-enhancing solution
    a. Na_2CO_3, 70.5 g
    b. H_2O, 10 liters

D. Procedure

1. Use glass pyrex dishes, 3-quart (12 × 17 × 3 inches) to stain gels. This size dish will guarantee adequate solution shaking, which is critical throughout the procedure. The efficiency of shaking is also

critical and should be optimized so that fluid recirculates efficiently without breaking gels.

2. Fix gels in gel fixative solution. Use gel volume–solution volume of 1 : 5.5. Up to four gels can be stained in trays, but *ratios of gel volumes to reagent volumes are critical throughout the procedure.* If you are staining only one thin gel, you may not have sufficient volume to easily equilibrate the gel. It is best that you stain multiple gels if this is the case.

3. For optimal results, gels should be fixed overnight. The following day, gels are washed in three changes of water for 1 hr each.

4. Incubate gels in silver stock solution for 1 hr. Use gel volume–solution volume ratio of 1 : 3. (Example, if gel is 50 ml, 150 ml silver stock should be used per gel).

5. Rinse gels briefly but thoroughly (10–15 sec) by suspending in water to remove surface silver. If they are not properly washed, black grains will form on the surface of the gels.

6. Add the reducing solution to the gels (gel volume–solution 1 : 5.5) while they are on the shaking platform and shake for 8–10 min. Immediately before adding the solution to the gels, formaldehyde is added to this solution.

7. Pour off the reduction solution and add the color-enhancing solution (gel volume–solution volume, 1 : 5.5). Shake in this solution for 45 min to 1 hr. Change the solution and shake for a second hour. Finally, change a third time and shake the solution overnight. The colors will appear within the first hour after the addition of Na_2CO_3, but the washes are necessary for full color development and will keep the gels from swelling excessively. (*Note:* If gels swell, it is generally due to inadequate volumes or shaking efficiency.)

8. This stain should give a uniform yellow background which will not interfere with black and white photography with Pan-X film which is suitable for publication.

9. For color slides, Ektachrome film (daylight) can be used. Color contact prints of gels can easily be made as described in Appendix 15.

Note: If background of gels is too dark for your preference, you may use warm sodium hydroxide solution and incubate in reducing solution (step 6) for 2 to 3 min longer.

Methods for Deglycosylation of Glycoproteins

I. TRIFLUOROMETHANESULFONIC ACID METHOD OF DEGLYCOSYLATION

Karp, D. R., Atkinson, J. P., and Shreffler, D. C., 1982, Genetic variation in glycosylation of the fourth component of murine complement, *J. Biol. Chem.* **257**(13):7330–7335.

A. Reagents

1. Trifluoromethanesulfonic acid, TFMS (Aldrich)*
2. Anisole (Aldrich)
3. Pyridine (Aldrich)
4. Ether absolute diethyl ether, ethyl ether (Aldrich)
5. Ammonium bicarbonate (NH_4HCO_3) (Sigma)

B. Equipment and Supplies

1. Glass Reactivials (Pierce)
2. Lyophilizer
3. Microfuge
4. Dialysis tubing (6 mm dry cylindrical diameter, Spectra Por #6)
5. Microfuge tubes, 1.5 ml

*Manufacturers in parentheses.

C. Stock Solutions

1. Ether–pyridine (9 : 1)
   a. Ether, 9 ml
   b. Pyridine, 1 ml
2. Ammonium bicarbonate
   a. 0.1 M NH$_4$HCO$_3$, 7.9 g
   b. Bring to 1000 ml

D. Procedure

1. Lyophilize protein sample in Reactivial to *complete* dryness before carrying out reaction.
2. Add 60 µl TFMS to 1 mg protein in the Reactivial (the '.'FMS *must* be completely anhydrous; therefore, it is recommended that you use an unopened vial of reagent every time you carry out this reaction). Add 10 µl anisole for every reaction.
3. Carry out the reaction for 2 hours in an ice bath.
4. Carefully remove the reacted material to an microfuge tube (1–1.5 ml) and very slowly (drop by drop) add a 9 : 1 mixture of ether : pyridine (be very careful because the mixture will spatter as it is precipitating).
5. When 1–1.5 ml of ether–pyridine mixture has been added, centrifuge in a microfuge for 1–2 min.
6. Remove ether–pyridine and resuspend the pellet (protein and pyridinium salt of the acid) in 0.1 M ammonium bicarbonate.
7. Dialyze sample against 0.1 M ammonium bicarbonate overnight.
8. Microfuge sample and analyze pellet for protein and carbohydrate (e.g., SDS–PAGE, lectin blotting).

II. DEGLYCOSYLATION OF GLYCOPROTEINS WITH ENDO-β-N-ACETYLGLUCOSAMINIDASE F

Elder, J., and Alexander, S., 1982, Endo-β-N-acetylglucosaminidase F: endoglycosidase from *Flavobacterium meningosepticum* that cleaves both high mannose and complex glycoproteins, *Proc. Natl. Acad. Sci. USA* **79**:4540–4544.

A. Reagents

1. Endoglycosidase F (New England Nuclear)
2. Sodium phosphate ($NaH_2PO_4 \cdot H_2O$) (Sigma)
3. EDTA (Sigma)
4. β-Mercaptoethanol (BioRad)
5. Sodium dodecyl sulfate (SDS) (Sigma)
6. Nonidet P-40 (Accurate)

B. Equipment

Water bath (37°C)

C. Stock Solution: Solubilization Buffer (pH 6.1)

1. 100 mM $NaH_2PO_4 \cdot H_2O$, 1.38 g
2. 50 mM EDTA, 1.9 g
3. 1.0% β-mercaptoethanol, 1 ml
4. 0.1% SDS, 0.1 g
5. Bring to 100 ml with distilled H_2O

D. Procedure

1. Solubilize protein (10–50 μg) to be deglycosylated by heating in 50 μl solubilization buffer (2–5 min at 95°C). (For some proteins, you may not need the SDS or β-mercaptoethanol and may therefore skip this step.)
2. Add 2–5 μl Endo F. (Concentration of Endo F will depend on source batch, etc., so you may have to use multiple concentrations.)
3. Add 10 μl of a 10% Triton X-100.
4. Incubate 4 hr at 37°C.
5. Lyophilize if necessary.
6. Resolubilize protein sample for analysis by one- or two-dimensional PAGE using SDS-solubilization buffer described in Appendixes 2 and 5. (*Note:* Proteolysis may be a problem due to protease contaminants in enzyme preparation. The use of two-dimensional PAGE instead of one-dimensional analysis will allow you to better determine whether reduction in the molecular weight is due to deglycosylation or proteolysis.)

Appendix 11

Antibody Preparation, Detection, and Characterization Techniques

I. PROPER HANDLING OF RABBITS

Since improper handling of rabbits may result in injuries that will affect the long-term health of the immunized animal, Figures A11.1–A11.3 illustrate the proper handling. Regardless of the immunization or bleeding procedures, rabbits should never cry and should be receptive to handling. In rare instances, a rabbit will be aggressive. This is usually because of previous improper handling. *Never* trap such an animal with a broom or other implement, but gently cover its head and eyes with a cloth (old laboratory coat) before you handle it. With proper handling, most any rabbit will become docile, and the trauma to animals and investigators becomes minimal.

II. PREPARATION OF EMULSION FOR IMMUNIZATION

A. Reagents

1. Immunogen (suspended in saline buffer)
2. Complete Freund's adjuvant (GIBCO or DIFCO)

B. Equipment and Supplies

1. 3–5 cc Luer-lock syringe
2. Plastic tubing (Tygon Microbore tubing [Norton Plastics & Synthetics] Fisher catalog #14-170-15F; S-54-HL, 0.30 × 0.090)
3. 18-gauge needles
4. Connecting needle apparatus (18-gauge needles connected by plas-

Figure A11.1. Proper handling of rabbits. I. Reach for the rabbit with a slow movement. Firmly grip skin just behind ears. (NEVER grab on to the ears themselves.)

Figure A11.2. Proper handling of rabbits. II. Lift rabbit from cage by back of neck while supporting hips firmly with other hand. (The rabbit cannot kick if held in this position.)

Figure A11.3. Proper handling of rabbits. III. If you must transport the animal for distances between rooms, you can easily do this by tucking the rabbit's head under your arm to hide its eyes while supporting its hips.

tic tubing. (*Note:* The precise tubing size must be used or you may lose your sample; see Fig. A11.4.)

C. Procedure

1. The acrylamide-supported protein spot (2–3 Coomassie Blue stained spots equivalent to 10–200 μg protein) can be minced into small pieces and placed into a 3-ml Luer-lock syringe containing 1.5 ml water or buffer.
2. A small piece of tubing is then used to connect this syringe to a second Luer-lock syringe (without needles) and the gel is passed back and forth between the two syringes until it is broken into small, uniform pieces.
3. The tubing is then removed and two 18-gauge needles are connected to the syringes. (If soluble immunogen and not polyacrylamide gel pieces are to be used, start with this step.)
4. An 18-gauge needle is connected to one syringe and 1.5 ml of complete Freund's adjuvant is brought into this syringe. The single needle is removed, and the connecting-needle apparatus is securely fastened into the Luer lock.

5. The adjuvant is slowly drawn into the syringe and the syringes are connected by tubing (see Fig. A11.4).

6. Emulsify the immunogen and adjuvant mixture as shown in Figure A11.4 by passing the mixture back and forth between the syringes until the sample is thick. (*Note:* Some immunogen-adjuvant mixtures will be extremely viscous, while others will not be as thick. This should not effect the immunogenicity, although immunization is more difficult if the emulsion is thick because the fluid will run out of the syringe.)

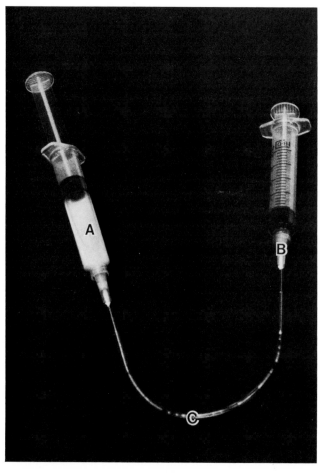

Figure A11.4. Syringe setup used for emulsifying acrylamide pieces with adjuvant. (A) 3-ml Luer-lock syringe. (B) 18-gauge needles. (C) Tygon tubing.

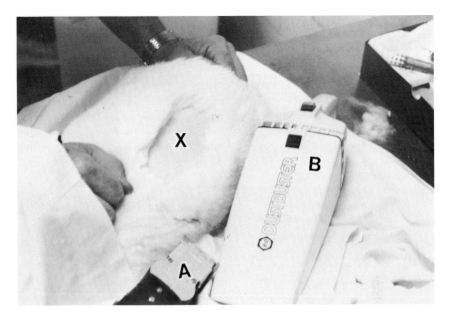

Figure A11.5. Preparation of rabbit for intradermal immunization. The hair on the upper sides of the rabbit (X) is shaved so that optimal injections can be given. (A) Electric razor. (B) Vacuum for hair.

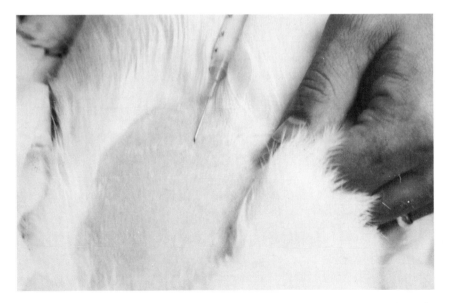

Figure A11.6. Intradermal immunization of rabbit. I. Note that the needle should go in parallel with the skin and should penetrate just into the layers of the skin (intradermal) and *not* beneath the (subcutaneous).

III. IMMUNIZATION OF RABBITS

A. Reagents

  1. Emulsion of immunogen–adjuvant
  2. 70% ethanol

B. Equipment and Supplies

  1. Small vacuum cleaner
  2. Electric shaver
  3. Luer-lock syringe (1–3 ml containing emulsion)
  4. 22-gauge needles

C. Procedure

  1. The emulsified mixture is injected into the layers of the skin of a shaved rabbit using a 22-gauge needle (see Figs. A11.5 and A11.6).

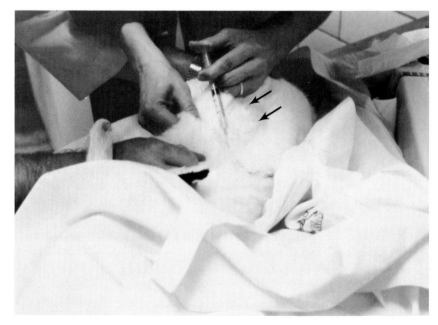

Figure A11.7. Intradermal immunization of rabbit. II. Small pockets of emulsion (arrows) should remain if injection is properly performed.

2. The needle is quickly inserted just under the surface of the skin and enough emulsion is added to form small raised "pockets" in the skin (Fig. A11.7).
3. A volume of approximately 0.5–1.0 ml should be injected into 10–12 sites along the upper sides of the rabbit. This region is chosen because the animal will be less likely to scratch or bite, and thus irritate, the sores that will occur when this adjuvant is used.
4. The remainder of the emulsion (0.5 ml) is also given subscapularly at this time. This is carried out by inserting the needle approximately ½ inch under the surface of the skin just between the shoulder blades. (*Note:* This procedure can be done with a minimum amount of stress to the animal.)
5. If you are immunizing a rabbit, simply sit it on an old lab coat and cover its eyes. If the animal cries, you are not handling it properly.
6. After 3–4 weeks, the first boost injection is given subcutaneously. It is generally adequate to emulsify your immunogen in incomplete Freund's adjuvant (using a 1 : 1 vol immunogen to adjuvant ratio) and inject this into a single site subcutaneously. As before, the amount of your antigen will depend on its immunogenicity and availability. In general, we use an equivalent amount or one-half of the amount used for the initial injection (10–200 μg).

If your molecule is sufficiently immunogenic, you should have serum antibodies within 10–14 days following the first boost injection. It is common, however, that the denatured proteins from polyacrylamide gels do not give high antibody titers by this time. In these cases, we wait until the sores on the animals have healed and then give a second intradermal boost in six to eight sites using the immunogen emulsified with complete Freund's adjuvant. In most cases, we have found that these animals develop significant serum titers after this boost within 4 weeks.

If your molecule does not prove to be immunogenic, it may be necessary to conjugate it to a hapten or chemically modify this protein, as discussed in Chapter 8.

IV. BLEEDING RABBITS

A. Reagents

1. 70% ethanol
2. Xylene
3. Vaseline

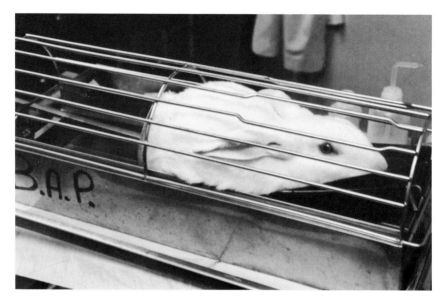

Figure A11.8. Restraining rabbit for bleeding. The best rabbit restrainer is one that does *not* trap the rabbit's neck since rabbits can easily be injured in cages where only the head is exposed. The head trough can be moved close to either side of the cage for bleeding or intravenous injection.

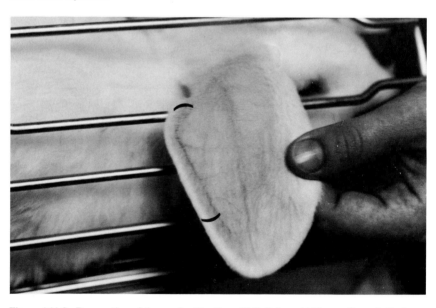

Figure A11.9. Preparation of the ear for bleeding. (1) Pull the rabbit's ear through the bars. (2) Trim the hair of the ear along the margin (brackets) using a single-edge razor blade. It is advisable to get the hair along the edge as well as on top of the vein. (3) Rinse the ear with 70% ethanol.

B. Equipment and Supplies

    1. Rabbit restrainer (#H3-240, Lithgow Laboratory Services, 1305 S. Railroad Avenue, San Mateo, CA 94402).
    2. Cotton swabs
    3. Single-edge stainless steel razor blades
    4. Glass blood-collecting vials
    5. Cotton balls
    6. Kimwipes
    7. Paper clips

C. Procedure

    1. Place rabbit in restrainer and push back of restrainer until animal is secured (Fig. A11.8). Place head trough to the same side as the ear to be bled.

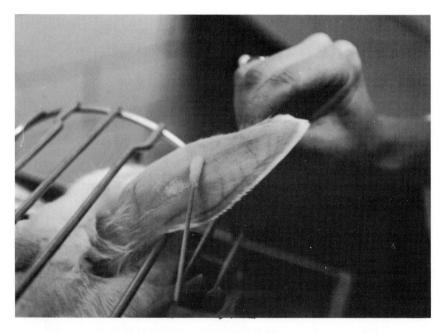

Figure A11.10. Application of xylene and Vaseline. (1) The use of xylene will cause vessel dilation and make it much easier to bleed. *Note:* Xylene can also damage if excessive quantities are used and if this is not washed off thoroughly. Apply xylene with a Q-tip on the center artery. *Never* apply on marginal vein. If too much xylene is used, the bleeding will also be difficult to stop. (2) Generously apply Vaseline on ear marginal edge and region just under ear. *Do not* apply where cut is to be made.

2. Prepare the ear for bleeding as shown in Figures A11.9 and A11.10.
3. Initiate bleeding and collect blood as illustrated in Figures A11.11 and A11.12. Blood should be collected in glass tubes and not plastic because clotting will be more efficient and more sera with minimal hemolysis can be obtained. Figure A11.13 summarizes the steps designed to stop the animal from bleeding.
4. The animal should stop bleeding immediately or very shortly after the ear is released.
5. As soon as the blood clots in the tube, rim the edges of the clot with a wooden applicator or glass pipette. (*Note:* This is a critical step for obtaining the maximum amount of serum from blood.)
6. Allow the blood to clot for 2–4 hr at room temperature; then decant sera and centrifuge at $1000 \times g$ for 10 min to remove blood cells. Some additional sera can be obtained by letting the blood clot overnight at 4°C, but there is usually more hemolysis in this sample.

Figure A11.11. Initiation of bleeding. Once the ear is prepared, a razor blade is used to cut into the vein. This is done using a firm, flat stroke perpendicular to the vein. A small parallel cut may also be used. Be careful not to cut through the ear! For a few seconds, the animal may not bleed, then a steady dripping of blood should start. If bleeding does not start, it may be necessary to gently recut to be sure that you have penetrated all layers of the vessel.

Figure A11.12. Collecting blood. Although the bleeding rate may vary slightly among animals, you should be able to easily collect 30–50 ml of blood in 5–10 min. The marginal vein should be held behind the cut (arrow) to increase pressure or to speed bleeding. The ear should also be bent slightly to open the cut and keep the blood from clotting. The blood should be collected into glass vials or beakers (not plastic).

7. Sera or purified immunoglobulins should be frozen in aliquots to avoid multiple freezing and thawing.

V. ENZYME-LINKED IMMUNOASSAY USING VECTOR LABORATORIES (VECTASTAIN) KIT: AVIDIN–BIOTINYLATED PEROXIDASE COMPLEX

Of all the enzyme-linked assays we have used during the past years, the avidin–biotinylated peroxidase complex (ABC) has proved to be one of the most sensitive and reproducible, therefore, it is one which we recommend. Although it is possible to conjugate enzymes to your own immunoglobulins, it is advisable that you use standardized reagents from commercial sources initially when you are establishing this technology in your laboratory.

A. Reagents

1. Sodium carbonate ($Na_2CO_3 \cdot H_2O$) (Sigma)
2. Sodium bicarbonate ($NaHCO_3$) (Fisher)
3. Sodium chloride (NaCl) (Sigma)
4. Sodium phosphate (monobasic) ($NaH_2PO_4 \cdot H_2O$) (Fisher)
5. Sodium phosphate (dibasic) ($Na_2HPO_4 \cdot 7H_2O$) (Fisher)
6. Tween 20 (Fisher)
7. Citric acid (Curtis Matheson)
8. Hydrogen peroxide (Sigma)
9. Methanol (Fisher)
10. Instant nonfat dry milk powder (Carnation) or bovine serum albumin (BSA) (Sigma)

Figure A11.13. Stopping the bleeding. (1) Hold the marginal vein just in front of the cut and bend the ear gently to close the cut. (2) Place a small amount of vaseline on a piece of cotton on the cut and hold firmly. (3) Release the back partition of the restrainer and allow the rabbit to relax. (4) Wash the ear with a small amount of alcohol and then thoroughly with a paper towel soaked well with cold water to remove xylene. A small amount of baby oil will also help the ear from drying out if multiple bleedings are to be carried out. (5) If bleeding persists, fold a Kimwipe and cover both sides of the ear around cut using a large paper clip to hold in place. This should be left in place for a few minutes but *never* forget and leave the rabbit with this on its ear.

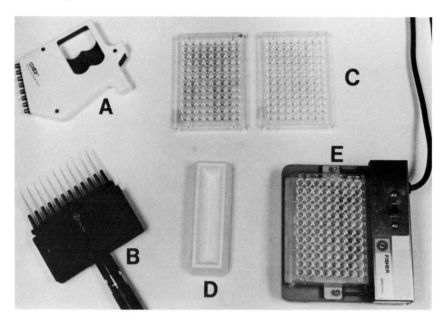

Figure A11.14. Equipment and supplies used in enzyme-linked immunoassays. (A,B) Micro-pipettors. (C) Microtiter plates. (D) Trough for reagents for micropipettors. (E) Vibrating shaker for assay incubation.

11. *O*-Phenylenediamine dihydrochloride (OPD) (Bethesda Research)

B. Equipment and Supplies (Fig. A11.14)

1. Microtiter plate (you may have to experiment to determine which brand of plates will optimally bind your antigen): We have found that the 96-well Immunlon 2 U plates (made by Dynatech and purchased through Fisher) brand plates have consistently given good results.
2. Multichannel micropipettes and trays (6 or 12 channel) (BioRad or Flow Laboratories)
3. Microtiter ELISA reader (Fig. A11.15). Although this assay can be carried out without an ELISA reader, most research and medical centers have these instruments. Because they will read dozens of assays within a few minutes, they are a valuable piece of equip-

ment for most studies using antibodies. Instruments are now available from Flow Laboratories, Dynatech Instruments, or BioRad.
4. Microtiter plate mixers, essential for assay sensitivity and reproducibility (Fisher, Flow Laboratories)
5. Vectastain ABC kit with biotinylated second antibody (Vector Laboratories)

C. Stock Solutions

1. Antigen-coating buffer (0.1 M sodium carbonate)
   a. 0.1 M $Na_2CO_3 \cdot H_2O$/100 ml H_2O, 1.24 g
   b. 0.1 M $NaHCO_3$/100 ml H_2O, 0.84 g
   c. Add (a) to (b) while monitoring the pH until pH 9.6 is reached. (Most antigens will bind adequately to microtiter plates when this pH is used. If your antigen does not bind, you may have to try different buffer systems or plates.)
2. Blocking solution (2% milk or BSA in antigen-coating buffer)
   a. 20 g instant nonfat dried milk or 20 g BSA

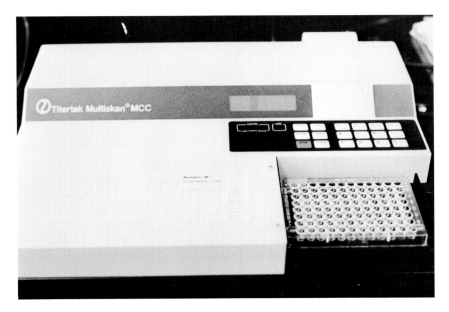

Figure A11.15. Example of type of equipment used to read enzyme-linked immunoassay in microtiter plates.

   b. Bring to 1000 ml with antigen coating buffer 1

   c. Stir well until milk is dissolved completely (solution may be aliquoted and stored frozen at $-20°C$ for later use).

     Note: The use of nonfat dry milk is an effective blocking agent and is considerably less expensive for use if large numbers of assays are to be processed. It is critical, however, to establish that the complex proteins in milk do not contain antigenic determinants that may cross-react with the antigen or antibody of interest.

3. Assay buffer (20 mM phosphate buffered saline plus Tween 20)

   a. 230 ml of a 0.02 M NaH_2PO_4, 2.78 g/liter

   b. 770 ml of a 0.02 M $Na_2HPO_4 \cdot 7H_2O$: 5.36 g/liter

   c. 150 mM NaCl, 8.18 g

   d. 0.05% Tween 20, 0.5 ml

   e. pH 7.3

     Note: Do not include sodium azide in your buffers, as this will affect enzyme activity.

4. Diluent for primary antibody (2% milk or BSA in assay buffer)

   a. 2% nonfat dried milk or BSA, 2 g

   b. Bring to 100 ml with assay buffer 3

   c. Store at $-20°C$

5. Substrate buffer (citrate/phosphate, pH 5)

   a. 0.05 M citric acid/50 ml distilled H_2O, 0.48 g

   b. 0.1 M $Na_2HPO_4 \cdot 7H_2O$/100 ml H_2O, 2.68 g

   c. Mix 24.3 ml of citric acid (solution a) with 25.7 ml of Na_2HPO_4 (solution b).

   d. Adjust pH to 5.0. with either solution a or b

   e. Store at $-20°C$

6. Primary and secondary antibody solutions (to be made on day assay is run)

   a. Dilute primary antibody in diluent buffer 4 (depending on antibody, the dilutions can range from 1 : 10^2 to 1 : 10^6). Because the reaction employed in this assay is extremely sensitive, you may have background readings of nonimmune sera at dilutions up to 1 $\times$ 10^3. It is therefore recommended that control nonimmune serum dilutions be used which are identical to immune sera be used initially to standardize assay.

   b. Dilute biotinylated second antibody in assay buffer 3. Dilution is usually 1 : 200.

7. Inactivating reagent (to be made fresh on day assay is run)

   a. 100 μl 30% H_2O_2 in 10 ml MeOH

8. Avidin–biotinylated peroxidase complex (ABC reagent) (dilutions made fresh on day of assay)
    a. Dilute reagent A 1 : 100 in assay buffer 3
    b. Dilute reagent B 1 : 100 in assay buffer 3
    c. Mix equal parts of solutions a and b
    d. Preincubate at room temperature for at least 1 hr before use
9. Enzyme substrate solution (made fresh on day of assay)
    a. Dissolve 8 mg substrate (OPD) in 10 ml substrate buffer 5
    b. Add 100 μl H_2O_2 to 10 ml of this substrate mixture
       Note: Use gloves to handle this carcinogenic substrate. Protect this substrate from light before use.

D. Procedure

1. Coat microtiter plates with antigen. Dilute antigen in antigen coating buffer (1). We have had good results with concentrations of 0.1–1 ng/μl, but this concentration will depend on the purity of your antigen as well as on the type of antibodies for which you are screening.
   Note: Controls for each assay should include (a) no antigen adsorbed to wells, (b) no primary antibody, (c) nonimmune serum (or immune sera from animals injected with adjuvant alone), and (d) no secondary antibody.
2. Add 50 μl antigen in coating buffer per microtiter plate well. Incubate at room temperature (25°C) for 6 hr or overnight at 4°C. (Use microtiter shaking apparatus for optimal results. In some instances, plates can be coated and then frozen for later use.)
3. After coating, wash plates thoroughly four times with assay buffer (3). (It is optimal to use a microtiter plate washer so if this is not available, be sure to wash thoroughly). After each wash, shake out contents sharply and thoroughly to prevent mixing of reagents between well.
4. Block plates by adding 100 μl/well blocking solution (2). (We have had optimal results by incubating overnight at 4°C with shaking, but shorter blocking times may be adequate for your needs if a qualitative rather than a quantitative assay is needed.) Wash plates 4× with assay buffer.
5. Apply primary antibody dilutions to microtiter wells (50 μl/well). (For quantitative purposes, we use 16 hr incubations overnight

at 4°C. For rapid antibody detection assays, this incubation can be reduced to a few hours at room temperature.)

6. Remove antibody from wells. Wash plates four times with assay buffer. Again, be careful that reagents do not mix between wells during washes.

7. Add second antibody (50 μl/well) and incubate for one hour at room temperature while shaking. Wash plates four times with assay buffer.

8. Add inactivating reagent (100 μl/well). Incubate at room temperature with shaking. Wash plates four times with assay buffer.

9. One hr after preparing ABC reagent (8), add 50 μl/well. Incubate for 1 hr at room temperature with shaking. Wash plates four times with assay buffer.

10. Add OPD substrate solution (9) (50 μl/well) and incubate for 30 min at room temperature with shaking.

11. Read microtiter plate using ELISA reader at 450 nm absorbance. Reaction cannot be stopped. Even control wells will be yellow after a few hours, so be consistent between assays as to when the plates are read.

VI. IMMUNOELECTROPHORESIS

Modified from Weeke, B., 1975, *A Manual of Quantitative Immunoelectrophoresis,* Universittsforlaget, Oslo; also 1973, Equipment, procedures, and basic methods, *Scand. J. Immunol.* (Suppl. 1) **2**:15–83.

A. Reagents

1. Sodium barbital (Mallinckrodt)
2. Barbital (Mallinckrodt)
3. Sodium azide (Sigma)
4. Triton X-100 (BioRad)
5. Agarose, electrophoresis quality (BioRad)
6. Trizma base (Sigma)
7. Glycine (Sigma)
8. Coomassie blue R-250 (BioRad)
9. 95% ethanol
10. Glacial acetic acid

B. Equipment and Supplies (Fig. A11.16)

    1. Electrophoresis apparatus (LKB multiphor or equivalent) with cooling plate and antidecondensation lid
    2. Power supply: 70–300 V, 150-mA capacity
    3. Gel puncher available commercially from LKB or BioRad (or use a small cork borer or glass Pasteur pipette attached to side-armed flask attached to a vacuum source)
    4. Hole templates (several patterns are commercially available from LKB, or BioRad)
    5. Water bath (45°–55°C)
    6. Large straight-edged spatula and razor blades
    7. 10–15-ml glass test tubes
    8. 100 ml Erlenmeyer flasks
    9. Filter paper wicks (size will depend on electrophoresis apparatus to be used)

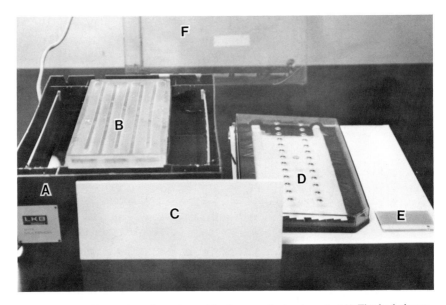

Figure A11.16. Electrophoresis equipment for immunoelectrophoresis. (A) Flat bed electrophoresis chamber (LKB multiphor unit). (B) Glass cooling plate. (C) Filter paper wicks. (D) Antidecondensation lid. (E) Agarose gel. (F) Lid for electrophoresis chamber.

10. Filtering funnel and Whatman #3 filter paper
11. Gel Bond (Marine Colloids Division, FMC Corp.)

C. Stock Solutions

1. Barbital buffer (concentrated)
   a. 0.1 M sodium barbital, 20.6 g
   b. 0.02 M barbital, 4 g
   c. 0.1% sodium azide, 1 g
   Dissolve the barbitals in 200–300 ml boiling distilled water. Allow to cool; then add sodium azide and bring up to 1 liter. Store at 4°C until use. Dilute 1 vol buffer with 4 vol distilled water prior to use.
2. Tris-glycine buffer (alternate buffer to barbital)
   a. Combine 56 g glycine with 45 g Trizma base and bring to 1000 ml with distilled water.
   b. Combine 13 g sodium barbital with 2 g barbital and bring to 1000 ml with distilled water.

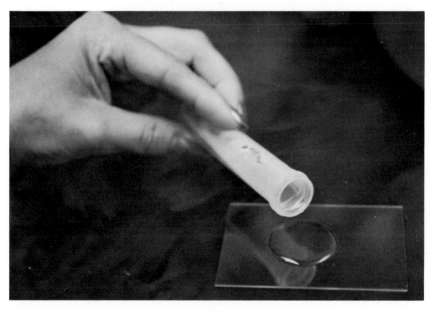

Figure A11.17. Preparing agarose gels for immunoelectrophoresis. Pour agarose (with or without antibody) onto slide until it uniformly covers the glass plate.

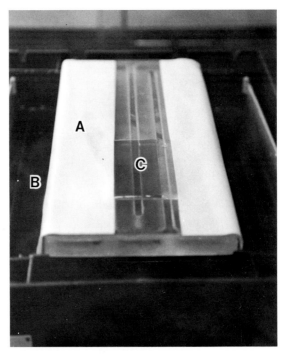

Figure A11.18. Preparing an agarose gel for electrophoresis. Place wicks (A) that have been wetted with buffer (B) in either chamber. Make sure there is a uniform contact of the agarose gel (C) with the filter paper.

    c. Mix equal volumes of solutions a and b and pH to 8.8. (This buffer has a higher buffering capacity and low ionic strength resulting in minimal pH changes during electrophoresis. It further does not require large quantities of barbital, which are controlled substances.)

3. Agarose
    a. 1% agarose, 5 g
    b. Barbital buffer, 500 ml
Heat buffer in boiling water bath. While stirring with a magnetic stir bar, slowly add agarose and heat until agarose is dissolved. (This may also be prepared using a microwave oven.) Aliquot into 15–50-ml plastic conical centrifuge tubes and store at 4°C until use.

4. Staining solution
    a. 0.5% Coomassie Brilliant Blue R-250, 5 g

    b. 45% ethanol, 450 ml

    c. 10% glacial acetic acid, 100 ml

    d. 45% distilled H_2O, 450 ml

    e. Make up stain and allow to sit overnight at room temperature before filtering

  5. Destaining solution

    a. 45% ethanol, 450 ml

    b. 10% glacial acetic acid, 100 ml

    c. 45% H_2O, 450 ml

  6. 0.1 M sodium chloride

    a. 0.1 M NaCl, 5.8 g

    b. Bring to 100 ml with distilled H_2O

D. Procedure for Immunoelectrophoresis

  1. Heat aliquot of agarose in boiling water bath or in microwave oven. Be sure that agarose is uniformly in solution.

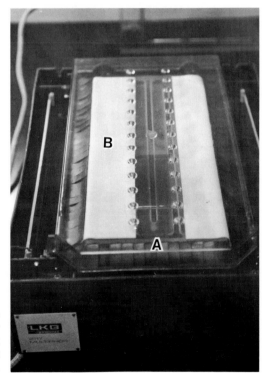

Figure A11.19. Place the antidecondensation lid (A) over the filter paper (B). This lid is important in that it prevents the agarose gel from drying out during electrophoresis.

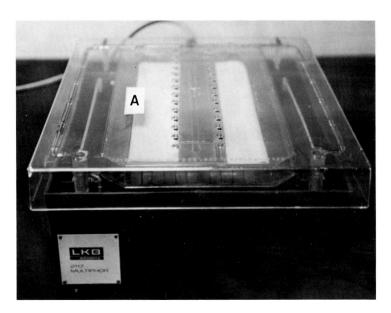

Figure A11.20. Place the lid (A) on the electrophoresis chamber, connect the electrodes to a power supply, and begin electrophoresis.

2. If antibodies or other proteins are to be added, allow the agarose to cool; just before it solidifies, add the antisera to the agarose and vortex thoroughly.
3. Pour the agarose evenly onto the glass slide or plate (as illustrated in Fig. A11.17). Make sure that surface is level in order to obtain uniform layer of agarose.
4. Punch wells in the agarose and place it in the chamber. Cover the edges of the gel with filter paper that has been soaked in buffer (Fig. A11.18). Prepare chamber for electrophoresis as shown in Figs. A11.18–A11.20.
5. *Note:* If rocket immunoelectrophoresis is to be used, antibodies will be added to the agarose. Figure A11.21 illustrates the experimental design to be used to initially determine if precipitating antibodies are to be used. If crossed or crossed-line electrophoresis are to be used, the antigen will be separated first by using agarose without antibody. The agarose containing antibody and the protein for line electrophoresis will then be poured adjacent to the first-dimension agarose (see Fig. A11.22 and discussion in Chapter 13).

6. Electrophoresis is carried out as follows:
   a. First-dimension electrophoresis for crossed immunoelectro-
      phoresis, etc. can be carried out at 10 V/cm for 2 hr.
   b. Second-dimension electrophoresis is carried out for 12–16 hrs
      at 2 V/cm.

E. Procedure for Drying and Staining Agarose Gels with Immunoprecipitates

1. Remove agarose gel from electrophoresis chamber. Place glass
   plate on paper towels and pour 0.1 M NaCl solution into antigen
   wells and over surface of agarose.
2. Place wet filter paper (two to three sheets) over agarose gel. Be
   sure to remove air bubbles.
3. Place several layers of paper towels on top of filter paper and
   weight these down (two to three large scientific supply catalogues
   are excellent for this).

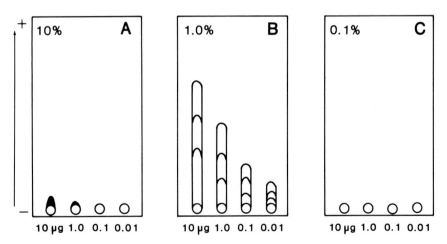

Figure A11.21. Example of experimental design of rocket immunoelectrophoresis to determine optimal ratios of antibodies and antigens for other immunoelectrophoresis methods. Slides with agarose containing different percentages of antisera (10, 1, and 0.1%). All slides contain wells with varying amounts of antigen (e.g., 10 μg, 1 μg, 0.1 μg, 0.01 μg). (A) Excess antibody results in precipitate formation near antigenic origin. (B) Adequate ratio of antigen to antibody for resolution of immunoprecipitates, showing the optimum ratios to be used for further immunoelectrophoresis studies.

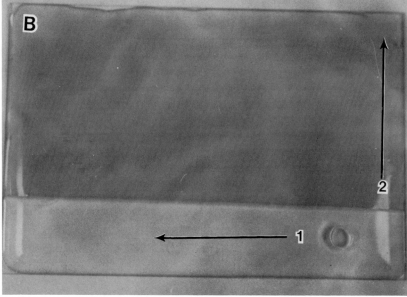

Figure A11.22. Preparing agarose gel after crossed immunoelectrophoresis. (A) Following separation of antigen by electrophoresis in the first dimension (1), cut agarose with a flat-edged knife and remove the agarose which does not contain the antigen. (B) Pour agarose containing antibody so that it is adjacent to first dimension before second-dimension electrophoresis (2) is carried out.

4. After 1–2 hr, remove filter paper and place gel in container containing 0.1 M NaCl solution. Shake *gently* for 1–2 hr. (If excessive amounts of hemoglobin are present in sera, you may need to wash gel overnight in multiple changes of NaCl solution.) Frequently, it is advantageous to remove the gel from the glass plate during the washing procedure.
5. After washing, place the agarose gel on a Gel Band and dry as described in #3 above. (*Note:* If the filter paper sticks to the gel, wet it with some water and it should immediately come off.)
6. Allow the gel to air dry or use a hair dryer to dry it. (The gel *must* be completely air dried before staining.)
7. Place gel in Coomassie Blue staining solution for 10–30 min.
8. Remove gel from staining solution and rinse gently with water to remove excess stain.
9. Place gel in destain solution until background is destained and immunoprecipitate is visible.
10. Allow gel to air dry.

VII. PARTIAL PURIFICATION OF IMMUNOGLOBULINS BY AMMONIUM SULFATE FRACTIONATION

A. Reagents

1. Ammonium sulfate [$(NH_4)_2SO_4$] (Sigma)
2. Serum or ascites fluid

B. Stock Solution: Saturated Ammonium Sulfate

1. $(NH_4)_2SO_4$, 777 g
2. Bring to 1000 ml with distilled H_2O
3. pH to 7.2–7.4 (*Note:* this is critical, as pH extremes may alter antibody activity.)

C. Procedure

1. Place serum in flask or beaker on a magnetic stir plate and begin mixing.
2. Slowly pour saturated $(NH_4)_2SO_4$ into serum while stirring. Add approximately 40 ml per 60 ml sera.

3. Allow precipitation to occur for 1 hr while stirring.
4. Centrifuge serum at 2000–5000 × g for 20–30 min.
5. Resuspend precipitated pellet with distilled H_2O. It is usually preferable to use as little volume as is possible to dissolve precipitate.
6. Dialyze fractionated sample extensively to remove excess ammonium sulfate. (If you are going to purify IgG by DEAE chromatography, dialyze against 0.02 M phosphate buffer.)

D. Stock Solutions

1. Monobasic phosphate, 0.02 M
   a. 0.02 M $NaH_2PO_4 \cdot H_2O$, 2.76 g
   b. Bring to 1000 ml with distilled H_2O
2. Dibasic phosphate, 0.02 M
   a. 0.02 M NaH_2PO_4, 2.84 g
   b. Bring to 1000 ml with distilled H_2O
3. Phosphate buffer, 0.02 M
   a. While monitoring the pH, add enough dibasic phosphate solution to 1 liter of monobasic phosphate solution until the pH is 7.4.
4. Monobasic phosphate, 0.2 M
   a. 0.2 M $NaH_2PO_4 \cdot H_2O$, 5.5 g
   b. Bring to 200 ml with distilled H_2O
5. Dibasic phosphate, 0.2 M
   a. 0.2 M NaH_2PO_4, 5.7 g
   b. Bring to 200 ml with distilled H_2O
6. Repeat 3a using 0.2 M solutions

VIII. DEAE–SEPHACEL ION-EXCHANGE CHROMATOGRAPHY FOR ISOLATION OF IgG

A. Reagents

1. DEAE–Sephacel (Pharmacia)
2. Sodium phosphate, monobasic, $NaH_2PO_4 \cdot H_2O$ (Fisher)
3. Sodium phosphate, dibasic, Na_2HPO_4 (Fisher)
4. Sodium chloride, NaCl (Sigma)

B. Equipment and Supplies

1. Column either 1.6 × 20 cm or 2.6 × 40 cm (Pharmacia)
2. Fraction collector (LKB, Pharmacia, or Gilson)
3. Plastic tubes to hold at least 10 ml (Sarstedt)
4. UV monitor—ISCO or spectrophotometer (Bausch & Lomb or Beckman) to read adsorbance at 280 nm
5. Dialysis tubing (<10,000-M_r cutoff) (Spectrapor from Fisher)
6. pH meter

C. Stock Solutions

1. Phosphate equilibration buffer (0.02 M) monobasic
   a. 0.02 M $NaH_2PO_4 \cdot H_2O$, 2.76 g
   b. Bring up to 1 liter with distilled H_2O
2. Phosphate equilibration buffer (0.02 M) dibasic
   a. 0.02 M NaH_2PO_4, 2.84 g
   b. Bring up to 1 liter with distilled H_2O
3. While stirring with pH meter in place, slowly add enough (2) dibasic to 1 liter of (1) monobasic to make the pH 7.4. This is the 0.02 M phosphate equilibration buffer.
4. Phosphate regeneration buffer (0.2 M) monobasic
   a. 0.2 M $NaH_2PO_4 \cdot H_2O$, 5.5 g
   b. Bring to 200 ml with H_2O
5. Phosphate regeneration buffer (0.2 M) dibasic
   a. 0.2 M NaH_2PO_4, 5.7 g
   b. Bring to 200 ml with H_2O
6. Repeat step #3 using 0.2 M solutions.
7. Various NaCl concentrations in phosphate equilibration buffer.
   a. 0.05 M NaCl, 1.46 g: Bring up to 500 ml in 0.02 M phosphate buffer.
   b. 0.1 M NaCl, 2.92 g: Bring up to 500 ml in 0.02 M phosphate buffer.
   c. 0.2 M NaCl, 5.84 g: Bring up to 500 ml in 0.02 M phosphate buffer.
   d. 0.5 M NaCl, 14.6 g: Bring up to 500 ml in 0.02 M phosphate buffer.
   e. 1.0 M NaCl, 29.2 g: Bring up to 500 ml in 0.02 M phosphate buffer.

D. Procedure

1. Dialyze sample (e.g., serum that has been ammonium sulfate precipitated) against the phosphate equilibration buffer at 4°C in tubing with a molecular-weight cutoff of less than 10,000 with at least three changes of buffer, generally overnight.
2. Pour the resin as a slurry into the column and let it settle. The height of the packed resin should be at least 2 cm below the top of the column. Rinse the column with phosphate buffer and make sure everything is flowing correctly.
3. Equilibrate the DEAE–Sephacel resin with the phosphate buffer.
4. Apply the dialyzed sample to the top of the column carefully and start collecting 8-ml fractions with the fraction collector.
5. Keep rinsing the column with phosphate buffer and analyzing the fractions by UV monitor or spectrophotometer (read the absorbance at 280 nm). The first IgG peak should come off in tube 8–10 of the 1.6 × 26 cm column. This column will have a bed volume of approximately 60 ml and can accommodate approximately 150–200 mg serum proteins. Keep rinsing until the absorbance goes down to background.
6. Then apply the 0.05 M NaCl in phosphate buffer until a peak comes off and the reading goes back down.
7. Repeat with the 0.1 M, 0.2 M, 0.5 M, and finally, 1 M NaCl in phosphate buffer to get all the protein off the column.
8. Regenerate the column by washing with at least three bed volumes of 0.2 M phosphate regeneration buffer. Equilibrate the column again with the 0.02 M phosphate equilibration buffer for storage.

IX. IODINATION OF IMMUNOGLOBULINS USING IODOGEN

Fraker, P. J., and Speck, J. C., Jr., 1978, Protein and cell membrane iodinations with a sparingly soluble chloroamide, 1,3,4,6-Tetra-chloro-3a,6a-diphenyl-glycoluril, *Biochem. Biophys. Res. Commun.* **80**:849–857.

A. Reagents

1. Iodogen (Pierce)
2. Sephadex G-75 (Pharmacia)
3. Sodium phosphate monobasic ($NaH_2PO_4 \cdot H_2O$) (Sigma)
4. Sodium phosphate dibasic ($Na_2HPO_4 \cdot 7H_2O$) (Sigma)

5. Chloroform (Fisher)
6. ^{125}I-carrier-free iodine, low pH, high concentration (New England Nuclear)

B. Stock Solution: 0.05 M Phosphate Buffer Stock

1. $Na_2HPO_4 \cdot 7H_2O$ (F.W 268) 13.4 g/liter H_2O
2. $NaH_2PO_4 \cdot H_2O$ (F.W 138) 6.9 g/liter H_2O

Add 202 ml solution 1 to 48 ml solution 2; pH should be approximately 7.3–7.5.

C. Equipment and Supplies

Radiation Safety Office approved-radioactive hood. (*Note:* ^{125}I is volatile; therefore adequate ventilation is essential.)

D. Procedure

1. Preparation of reaction vial
   a. Weigh out 20 mg Iodogen. Dissolve in 2 ml chloroform. Dilute 1 : 100 to give 0.1 mg/ml. Add 100 μl of this 0.1 mg/ml Iodogen/chloroform to a 12 × 75 mm glass test tube. Dry down with a gentle N_2 stream.
   b. Preparation of Sephadex column: Swell 1–2 g Sephadex G-75 in PBS, pH 7.4. Degas swelled Sephadex and pour a column approximately 0.7–1.0 × 18–22 cm. Equilibrate the column with at least three bed volumes of phosphate buffer. Block nonspecific sites with 1–3 ml 1% BSA in phosphate buffer. Wash out excess BSA with three more bed volumes of phosphate buffer. (This much can be done a day or two ahead and the column stored in the cold room or refrigerator. Do not allow the column to dry out at any time!)
2. *Use gloves and proper clothing as well as radioactive hood throughout this procedure.* To begin the iodination reaction, add the protein solution to be iodinated (50–100 μl) to the prepared reaction vial. Add 1 mCi ^{125}I (usually diluted with phosphate buffer to insure proper measuring). Gently mix so the protein solution, ^{125}I, and Iodogen will react. Allow to incubate 5–10 min (depending

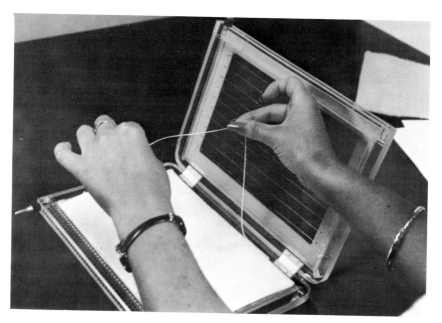

Figure A12.3. Immunoblotting procedure III. While laying polyacrylamide gel on filter paper, it is critical that no air bubbles be present between the paper and the gel.

Figure A12.4. Immunoblotting procedure IV. Carefully place wet nitrocellulose paper on top of the gel. Be careful not to handle paper more than necessary.

Figure A12.5. Immunoblotting procedure V. Place two sheets of wet filter paper on top of nitrocellulose.

Figure A12.6. Immunoblotting procedure VI. Add foam pad, making sure that it is sufficiently thick to ensure that the polyacrylamide gel is tightly compressed against the filter paper.

Figure A12.7. Immunoblotting procedure VII. Close filter unit and place into electrotransfer apparatus.

2. Preparation of gel for transfer: The two-dimensional-PAGE gel and the nitrocellulose paper are sandwiched between layers of filter paper and foam pads (see diagrams in Figs. A12.1–A12.7). Be careful not to allow any air bubbles between the nitrocellulose paper and the gel. It is possible to transfer two to three gels at a time using the commercial units described above. Every layer of filter paper, gel, and nitrocellulose should be wetted with buffer while constructing the sandwich. The gels are placed into the electrophoresis chamber so that the nitrocellulose paper is on the anode side of the gel. The final set up for your transfer system is diagramed in Figure A12.8.

3. Electrophoresis. If you do not use methanol in your electrophoresis buffer, all your proteins should be coated with SDS. Therefore, they will maintain their negative charge so that they will begin to move out of the acrylamide gel onto the paper immediately. We routinely start with chilled electrode buffer and carry out the electrophoresis for 2.5–3 hr. Amperage should be held at 1.0–1.5 A to prevent heating. (If a longer time is necessary for your proteins, you may have to use an efficient cooling buffer recirculating unit.) The recirculating and cooling, along with the

high ionic strength buffer, will help prevent the gel from swelling during transfer, a property that causes poor resolution of proteins on nitrocellulose paper. Remove the nitrocellulose paper and put in a glass pan with the protein transfer side (the one next to the gel) facing up and carry out blotting procedure.

II. ANTIBODY DETECTION OF PROTEINS TRANSFERRED TO NITROCELLULOSE USING ^{125}I PROTEIN A OR SECOND ANTIBODY

A. Reagents

1. Sodium chloride (NaCl) (Sigma)
2. Trizma base (Sigma)
3. Sodium azide (NaN$_3$) (Sigma)
4. Instant nonfat dry milk (Carnation) (local grocers) or bovine serum albumin (Sigma)

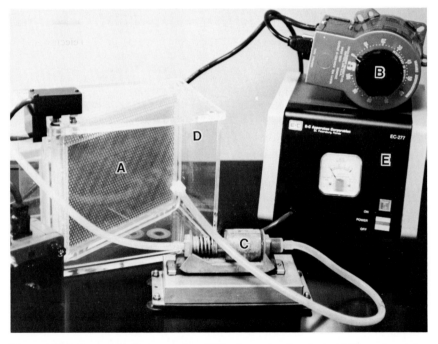

Figure A12.8. Immunoblotting procedure VIII. Apparatus for electroblotting of proteins (E-C Apparatus Corp.). (A) Filter unit with polyacrylamide gel and nitrocellulose paper. (B) Rheostat to control current generated by power supply. (C) Buffer recirculating pump. Note direction of flow from anode to cathode. (D) Electrophoresis chamber filled with electrode buffer. (E) Power supply.

5. Primary antibody
6. Second antibody intermediate (immunoglobulin which will react with primary antibody, e.g., if human sera is used, rabbit anti-human IgG should be used as the second antibody. Affinity-purified antibody is recommended). Because some species of IgG do not bind purified protein A (e.g., sheep or goat IgG), you will have to use labeled second antibody or a rabbit anti-sheep or goat antibody prior to addition of ^{125}I-labeled protein A (see discussion in Chapter 13).
7. [^{125}I]protein A (Amersham) OR
8. [^{125}I]-IgG directed against species from which primary antibody is obtained.

B. Equipment and Supplies

1. Glass pans that are slightly larger than the nitrocellulose (if you use plastic dishes, it is not advisable to reuse them, since many types of plastics will absorb reagents).
2. Shaking Platform (Health Products; Eberbach)

C. Stock Solutions

1. 10X stock Tris-buffered saline (pH 7.2–7.5)
   a. 9.0% NaCl, 90 g
   b. 0.10 M Trizma base, 12.1 g
   c. 0.2% sodium azide, 2 g
   d. Bring up to 1000 ml with H$_2$O
   e. Adjust pH to 7.2–7.5
2. Tris-buffered saline
   a. 100 ml, 10X stock Tris–saline buffer
   b. Dilute with 900 ml distilled H$_2$O
3. Blocking and serum dilution stock (2% milk in TBS)
   a. 2% instant nonfat milk, 20 g
   b. 1000 ml TBS
   c. Mix well to dissolve (solution may be stored frozen in aliquots at −4°C).

D. Procedure

1. Block nitrocellulose paper containing transferred protein as described with blocking buffer (3) for 6–24 hr while shaking vigor-

ously. Approximately 100 ml solution should be used to cover a 20 × 20 cm nitrocellulose square. (Be sure that solution uniformly covers nitrocellulose.)

2. Pour off blocking solution; add primary antibody diluted in blocking buffer. (Primary antibody can usually be added directly to blocking buffer if proper pH is used). Antibody dilution will depend on titer of antibody. When first testing new sera, we routinely dilute 0.2–2 ml sera (if available) in 80 ml blocking solution for a 20 cm^2 transfer. For best results, incubation is carried out for a minimum of 4–8 hr. (*Note:* Once the presence of antibody is established, you may be able to reduce both the amount of antibody as well as the time of incubation.)

3. Rinse off unbound antibody with Tris buffer. Allow the transfer to wash while shaking for 60 minutes using two changes of Tris buffer (2).

4. [^{125}I]protein A procedures
   a. If rabbit, human, or nonhuman primate, dog or cat sera are used, [^{125}I]protein A (10^6 cpm/50 ml:specific activity >30 mCi/mg can be added directly to blot and incubated overnight. If mouse, sheep, or goat sera are used, it will be necessary to use a rabbit anti-mouse, sheep, or goat immunoglobulin as an intermediate step because the Ig of these species does not adequately bind to purified protein A (see Discussion in Chapter 13).

5. [^{125}I]second antibody
   It is possible to iodinate second antibody (i.e., goat anti-rabbit IgG; rabbit anti-mouse IgG, etc.) directly as described in Appendix 11 (10^6 cpm/50 ml:specific activity is used for incubation).

6. The radiolabeled probes are removed from plates and nitrocellulose is washed thoroughly with Tris–saline buffer (two changes each) for 30 min with shaking.

7. Nitrocellulose papers are thoroughly air dried or dried with a hair dryer before radioautography is carried out (see details in Chapter 6 for choice of x-ray films and types of exposure).

III. IMMUNOBLOT METHOD USING AVIDIN–BIOTIN METHOD

A. Reagents

1. Sodium carbonate ($Na_2CO_3 \cdot H_2O$) (Sigma)
2. Sodium bicarbonate ($NaHCO_3$) (Sigma)

3. Sodium chloride (NaCl) (Sigma)
4. Hydrogen peroxide (H_2O_2) (Sigma)
5. Carnation instant nonfat dry milk powder or bovine serum albumin (BSA)
6. 3,3'-diaminobenzidine (DAB) (Sigma)
7. Vectastain kit (avidin–biotinylated peroxidase) (Vector)
8. Trizma base (Sigma)
9. Nickel chloride ($NiCl_2$) (Sigma)

B. Equipment

1. Nitrocellulose (BioRad)
2. 1- or 2-μl pipette or 5-μl Hamilton syringe

C. Buffers

1. Antigen-coating buffer
   a. 0.1 M $Na_2CO_3 \cdot H_2O$, 12.4 g/liter
   b. 0.1 M $NaHCO_3$, 8.4 g/liter
   c. Mix together while monitoring the pH to give pH 9.6
2. Antigen-coating solution
   a. Dilute antigen in antigen-coating buffer to give a range of 2 μg–100 pg/2 μl.
3. 10X Tris-buffered saline
   a. 0.1 M Trizma base, 12.1 g
   b. 9.0% sodium chloride, 90 g
   c. Fill to 1 liter, pH 7.2–7.4
   d. Dilute one part 10X TBS with 9 parts distilled H_2O to give 1 × TBS.
4. Blocking and incubation buffer
   a. 3% instant nonfat dry milk powder
   b. Or 3% BSA
   c. 3 g/100 ml 1X TBS
5. Substrate buffers
   a. 0.1 M Trizma base, pH 7.2–12.1 g/1 H_2O
   b. 0.02% hydrogen peroxide, 10 μl (30% hydrogen peroxide) in 15 ml H_2O
   c. 8.0% nickel chloride, 0.8 g/10 ml H_2O

6. Substrate solution
   a. Mix equal volumes Trizma base buffer containing 1 mg/ml DAB with 0.02% hydrogen peroxide.
   b. Add nickel chloride to give 0.04%. (*Example:* 10 ml 0.1 M Trizma base buffer, pH 7.2; 10 mg DAB; 10 ml 0.02% hydrogen peroxide; 100 μl 8.0% nickel chloride.)

D. Procedure

1. Dot 2-μl samples of antigen diluted in antigen coating buffer on nitrocellulose paper. Allow to air dry.
2. Block nitrocellulose with 20 ml (for 88 × 88 cm nitrocellulose) blocking buffer, i.e., TBS + 3% milk at least 2 hr or overnight, room temperature, shaking.
3. Wash with two changes 20 ml each TBS + milk 10 min each, shaking.
4. Dilute primary antibody 1 : 200 or 0.1 ml antibody : 20 ml TBS + milk 2 hr or overnight, room temperature, shaking.
5. Wash with two changes 20 ml each TBS-milk 10 min each, shaking.
6. Wash. Prepare A and B reagents from Vectastain kit. Dilute A 1 : 100 (0.1 : 10); dilute B 1 : 100 (0.1 : 10) mix equal volumes together and allow to incubate 1 hr before use.
7. Dilute second antibody from vectastain kit 1 : 200 or 0.1 ml antibody : 20 ml TBS + milk 1 hr, room temperature, shaking.
8. Add A,B solution (20 ml) 1 hr, room temperature, shaking.
9. Wash.
10. Add 20 ml substrate solution.

IV. LECTIN BLOTTING OF GLYCOPROTEINS TRANSFERRED TO NITROCELLULOSE

A. Reagents

1. Sodium chloride (NaCl) (Sigma)
2. Tris-base (Sigma)

3. Sodium azide (Sigma)
4. Bovine serum albumin, deglycosylated (Sigma)
5. Biotinylated lectin (Vector) (E-Y Laboratories)
6. 3,3'-diaminobenzidine (DAB) (Sigma)
7. Avidin-conjugated horseradish peroxidase (Vector)
8. 30% hydrogen peroxide (Sigma)
9. Nickel chloride (Sigma)

B. Equipment and Supplies

1. Glass pyrex pans or plates that are slightly larger than the nitrocellulose paper
2. Shaking platform

C. Stock Solutions

1. 10X stock Tris buffer (pH 7.2–7.5)
   a. NaCl, 90 g
   b. Tris-base, 12.1 g
   c. Adjust pH to 7.2–7.5.
2. IX Tris buffer
   a. 100 ml-10X stock Tris buffer
   b. Dilute with 900 ml distilled H_2O.
3. Nickel chloride (store solution in dark)
   a. $NiCl_2$, 0.8 g
   b. Bring to 10 ml with distilled H_2O.
4. Substrate solution (make fresh before using)
   a. Dilute 10 μl hydrogen peroxide in 15 ml distilled H_2O to make
      a 0.02% stock.
   b. Add 100 μl 8% $NiCl_2$ solution to 0.02% H_2O_2 stock.
   c. Add 1 mg/ml diaminobenzamide in Tris–HCl buffer and add to
      $NiCl_2$–H_2O_2 solution.

Electroelution of Proteins from Polyacrylamide Gels*

A. Equipment and Supplies

1. 5-ml glass pipets, Corning Pyrex brand
2. Dialysis tubing (6-mm dry cylindrical diameter; Spectra/Por #6, Spectrum Medical Industries) (Molecular-weight cutoff of tubing is optional.)
3. Dialysis clamps (Spectrum Medical Industries)
4. Razor blade or scalpel
5. Tweezer or spatula
6. Tube electrophoresis unit that will accommodate grommets for 5-mm OD tubes. *Example:* Hoefer GT3, GT5, GT6, or grommets UB3-G (Rubber corks with holes drilled in them may also be used if necessary.)
7. Surgical tubing, diameter to fit snugly over 5-ml disposable pipets, several inches
8. Ammonium bicarbonate (NH$_4$HCO$_3$)
9. Sodium dodecyl sulfate (SDS; BioRad)

B. Stock Solutions

1. Solution A
   a. 0.01 M ammonium bicarbonate, 1.6 g
   b. 0.1% SDS, 2.0 g
   c. Add distilled H$_2$O to 2 liters

---

*Modified from Braatz, J. A., and McIntire, K. R., 1977, A rapid and efficient method for the isolation of proteins from polyacrylamide gels, *Prep. Biochem.* **7**(6):495–509.

2. Solution B
   a. 0.01 M ammonium bicarbonate, 1.6 g
   b. Add distilled H_2O to 2 liters

C. Procedure

1. Preparation of dialysis tubing: Cut tubing into 6-inch lengths, rinse well in distilled H_2O, and soak in solution A. (Other types of dialysis tubing may have to be treated more extensively before use—consult manufacturer's instructions.)
2. Preparation of elution tubes: Slide one piece of prepared dialysis tubing over the tip end of a 5-ml pipette cut off at the 1.5-ml mark (pipette is now 5.5 inches long). It must fit snugly. Secure the dialysis tubing at least 1 inch above the tip end with a 5-mm section of surgical tubing. Clamp the open end of the dialysis bag. Fill the tube and dialysis bag with solution A, making sure there are *no* bubbles in the dialysis bag (see Fig. A13.1).
3. Mince the protein-containing gel pieces into 2-mm squares with a sharp razor blade or scalpel.
4. Use a tweezer or spatula to load the pieces into the buffer-filled tube, being careful to keep the dialysis tubing moist. Do not pack the tubes too tightly, as the gel will swell during elution. Place tubes into electrophoresis unit so that clamps just touch bottom of reservoir.
5. Fill the top and bottom reservoir of the electrophoresis unit with 0.01 M NH_4HCO_3 and 0.1% SDS. *Do not* allow lower buffer to come above the level of the surgical tubing.
6. Begin electroelution 10–15 mA (80–100 V, constant current). Continue at least 4–6 hr or until Coomassie Blue dye is tightly packed in the dialysis bag and a very sharp interface can be seen.
7. To begin electrodialysis to remove SDS, take out as much buffer as possible from inside the tube, open the dialysis clip, and allow the eluted protein to flow into a vial.
8. Reassemble clean 5-ml pipette tubes and fresh dialysis tubing. Load the eluted protein, overlay with 0.01 M NH_4HCO_3, and replace top and bottom buffer of the electrophoresis tank with 0.01 M NH_4HCO_3. Allow the electrodialysis to run at 10–15 mA (80–100 V) until the Coomassie has again packed down in the dialysis bag.
9. Carefully remove the eluted, dialyzed, and concentrated protein. This can be easily done without diluting sample by placing a dialysis tubing clamp above the protein interface so that the tubing can be cut just above the clamp (Fig. A13.1).

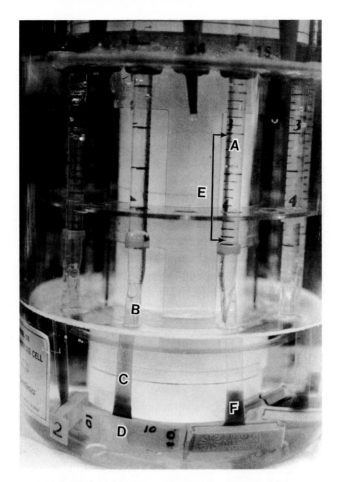

Figure A13.1. Apparatus used for electroelution of large quantities of proteins from poly-acrylamide gels (BioRad Laboratories). (A) Glass pipette. (B) Tip of pipette. (C) Dialysis tubing. (D) Dialysis tubing clamp. (E) Acrylamide pieces. (F) Electroeluted protein.

D. Modifications for Electroelution for Peptides for Amino Acid Sequence

Hunkapiller, M. W., Lujan, E., Ostrander, F., and Hood, L. E., 1983, Isolation of microgram quantities of proteins from polyacrylamide gels for amino acid sequence analysis, *Methods Enzymol.* **91**:227–236.

The following modifications have been used for polyacrylamide gel electrophoresis in which peptides generated by enzymatic digestion (e.g.,

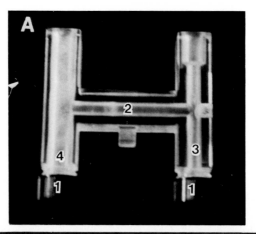

Figure A13.2. Electroelution device (Medical Specialties, MA) designed by Hunkapiller and Hood. (A) Individual electroelution chamber: 1, screwcap over dialysis tubing; 2, chamber connection for migration of protein; 3, acrylamide pieces; 4, concentrated electroeluted protein. (B) Tank for electroelution: 5, mirror reflection of electroelution chamber; 6, buffer overflow; 7, holes for tubing for buffer recirculation; 8, electroelution buffer.

trypsin, V-8, cyanogen bromide) are to be electroeluted from the gels for amino acid sequencing.

1. To prepare the sample for Laemmli electrophoresis or isoelectric focusing (IEF), heat for 10–15 min at 50°–60°C, rather than the more commonly used temperature of 95°C, in SDS solubilization buffer. The addition of sodium thioglycolate (0.1 mM) to the cathode (upper) buffer aids in scavenging destructive species of free radicals or oxidants trapped in the gel matrix by traveling at the dye front before the proteins. Side chain (tryptophan, histidine, and methionine) destruction can also be minimized by allowing the polymerized gel to sit overnight to clear it of free radicals.

2. Pour the gels 24 hr in advance to ensure removal of free radicals from TEMED and ammonium persulfate used to catalyze the acrylamide : bisacrylamide polymerization.

3. Electroelute using the elution chamber described by Hunkapiller *et al.* (1983), illustrated in Figure A13.2 (available from Medical Specialties, MA). This chamber has been designed to electroelute and electrodialyze very small quantities of proteins from polyacrylamide gels. *Note:* It is critical to recirculate the buffer during electroelution because the surface area of the dialysis tubing is limited.

Photography of Polyacrylamide Gels and Development of 35-mm Film

A. Reagents

1. Microdol X-developer (Kodak)
2. Rapid Fixer (Kodak)
3. Photoflo solution (Kodak)

B. Equipment and Supplies (see Figs. A14.1 and A14.2)

1. Ektachrome (color slide film)
2. Panatomic X film (black and white print film) (100-ft roll in individual rolls, Kodak)
3. Dark bag
4. Bulk film loader (Watson, model #100-35 mm, Pfefer Products, Simi Valley, CA)
5. Stainless steel film developing tank (Omega)
6. Plastic film apron
7. Film cassette for bulk loader
8. Camera stand (and 35-mm camera) (Polaroid camera or 4 × 5-inch film setups can also be used, but these are very expensive if large numbers of gels are to be processed.)
9. Light box
10. Stained polyacrylamide gel to be photographed

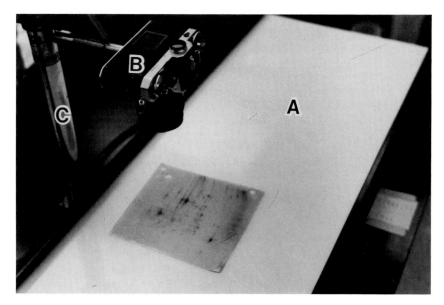

Figure A14.1. Setup for photographing polyacrylamide gels. (A) Light box. (B) 35-mm camera. (C) Camera stand.

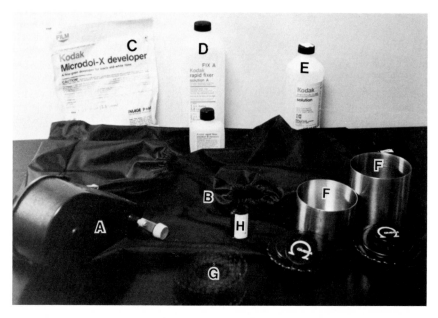

Figure A14.2. Equipment and supplies for developing black and white 35-mm film in the laboratory. (A) Bulk film loader. (B) Film cassette loading dark bag. (C) Microdol X-developer. (D) Rapid fixer. (E) Photoflo solution. (F) Film developing canisters with lightproof lid and cap. (G) Film apron. (H) Film cassette for loading film from bulk loader.

C. Stock Solutions

1. Microdol X-Developer

Slowly add contents of package (17 oz, 482 g) to 3.3 L H_2O and stir well until dissolved. (This will take a while if water is room temperature, but it may be easier than heating the water as the package suggests.) Bring volume up to 3.8 liters with H_2O.

2. Rapid Fixer

Add solution in bottle A (946 ml) to 2 liters room temperature distilled H_2O and stir. While stirring rapidly, add solution in bottle B (104 ml). Add water to bring volume to 3.8 liters.

3. Photoflo Solution

Dilute the Photoflo solution 1 : 200 by adding 1 ml to 199 ml H_2O.

D. Procedure for Photography of Gels

1. Load Panatomic X film from bulk film loader.
2. Load 35-mm camera.
3. Set up camera stand and mount camera as illustrated in Figure A14.1.
4. Focus camera on area of polyacrylamide gel of interest.
5. Use light meter on camera to determine approximate exposure. It is advisable to take bracket exposures by opening or shutting the f stop by two units to ensure that you will get an optimal exposure. We have found that with our light box, optimal exposures are obtained at f8, $\frac{1}{15}$ sec for panatomic and f8, $\frac{1}{8}$ sec for Ektachrome film.

E. Procedure for Developing Panatomic X Film

Once the film has been exposed, rewind it back into the cassette *inside* the camera. Put the cassette, a film apron, and a developing canister with its lightproof lid and small cap into a dark bag that is zipped up (Fig. A14.3). Undo the film cassette, take the end of the film, and wind it around

Figure A14.3. Loading film into developing canister from film cassette using dark bag. Both the outer and inner zippers of dark bag are unzipped. Developing canister, exposed film in cassette, and film apron are placed in dark bag and zippers are closed.

Figure A14.4. Loading the developing tank. The film is removed from the film cassette and is wound in the film apron.

the film apron in a nice tight circle so that the film never touches film (Fig. A14.4). Be sure to handle only the edges of the film. Put wound film with apron into the canister and tightly close the lightproof lid. Take everything out of the dark bag.

Develop film by adding Microdol X Developer through the hole in the lightproof lid. (Use 250 ml for the small one-film canister.) Put small cap on and agitate for 7 min. Pour off developer and fill canister with water. Immediately pour off and add rapid fixer. Agitate with cap on for 3 min. Pour off and rinse film for at least 10 min by taking off the lightproof lid and running tap water gently into the canister. Pour off water and soak film in diluted Photoflo ~30 sec. Wipe excess water off by running film through fingers and hang up to dry.

Direct Color Contact Printing from Polyacrylamide Gels*

A. Reagents

   Ilford Cibachrome-A Process P-30 Chemistry Kit

B. Equipment and Supplies (Fig. A15.1)

   1. Beseler Roller Tube (8 × 10″)
   2. Ilford Cibachrome-AII Deluxe Glossy Paper 8 × 10, CPSA.1K
   3. Darkroom with enlarger (color enlarger is optimal, otherwise a black and white enlarger and a set of filters)
   4. Unicolor Color Printing Filter Set (Unicolor Division, Photo Systems, Inc., Dexter, MI 48130)
   5. Contact Printing Frame 8 × 10 inches or larger

C. Stock Solutions

   Make developer; bleach and fix according to package insert (Ilford Cibachrome-A Process P-30 Chemistry Kit).

D. Procedure

   1. Place a wet silver-stained polyacrylamide gel on top of the contact printer frame (make sure excess fluid is removed from around gel). Be sure to align the gel with the acidic proteins on the left and

*Method modified from B. S. Dunbar.

the basic proteins on the right to follow the standardization of printing two-dimensional polyacrylamide gels. Gently remove air bubbles that may be trapped between glass.

2. Place the correct filter combination on the enlarger. A combination we have found acceptable for most contacts includes 0.4 cyanogen filters, 0.2 magenta filters using a 6-sec exposure with an f 5.6 aperture. The filter combination will vary among gels and light sources and may require trial and error.

3. In complete darkness (without safe light), slide a piece of Cibachrome paper (emulsion side up) into the correct place in the contact printing frame (under the glass plate). Cibachrome paper has emulsion on one side only. The instructions will tell you how to determine which side has the emulsion.

Figure A15.1. Equipment and supplies for color contact prints of polyacrylamide gels. (A) Roller tube for print development. (B) Plate for contact prints. (C) Ilford Cibachrome-A process P-30 chemistry kit. (D) Color printing filter kit (not necessary if filters available in color print enlarger). (E) Ilford Cibachrome-AII paper. (F) Photograph of gel.

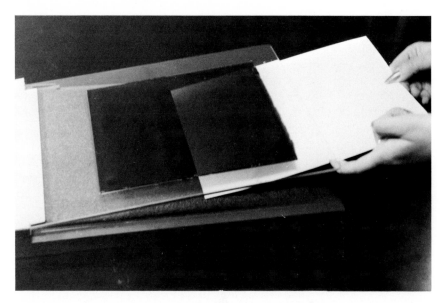

Figure A15.2. Setup for exposure of color print. Gel is placed on top of glass and paper is placed underneath. (*Note:* this should be carried out in darkness.)

4. Expose the paper for the proper time. (Again, experimentation using your light source will be the only way to determine the exposure time for your gel.) (See Fig. A15.2.)
5. Remove the exposed paper from the contact printing frame and roll to fit into the Beseler Roller Tube with the emulsion side of the paper facing the inside of the tube. More complete diagram and directions will be included with your roller tube.
6. Process the paper according to the Chemistry Kit directions. This usually requires 3 min in each solution developer, bleach and fix with a 10-sec wash between each. Volumes are discussed in Chemistry Kit instructions. After the roller tube is *closed,* the darkroom lights can be turned up to process the paper.
7. Wash and dry prints according to Cibachrome Paper AII instructions.

Addresses of Major Equipment Suppliers

Accurate Chemical and Scientific
Corporation
300 Shames Drive
Westbury, NY 11590
(516) 433-4900

American Scientific
100 Raritan Center Parkway
Edison, NJ 08818
(201) 494-4000

Amersham Corporation
2636 South Clearbrook Drive
Arlington Heights, IL 60005
(312) 593-6300

Amicon Corporation
17 Cherry Hill Drive
Danvers, MA 01923

Bethesda Research Laboratories
8400 Helgerman Court
Gaithersburg, MD 20877
(301) 258-8263

Bethyl Laboratories, Inc.
P.O. Drawer 821
Montgomery, TX 77356

BioImage Corporation
P.O. Box 2327
333 Parkland Plaza
Ann Arbor, MI 48106
(313) 994-6363

BioRad Laboratories
2200 Wright Avenue
Richmond, CA 94804
(415) 234-4130
(800) 227-5589 (outside CA)
(800) 227-3259 (inside CA)
(800) 645-3227 (outside NY)
(800) 632-3060 (inside NY)

Boehringer Mannheim Biochemicals
7941 Castleway Drive
P.O. Box 50816
Indianapolis, IN 46250
Telex 272330
(317) 849-9350

Calbiochem-Behring Brand Biochemicals
10933 North Torrey Pines Road
La Jolla, CA 92037
(519) 450-5680
(800) 854-9256
P.O. Box 12087
San Diego, CA 92112

Cappel/Worthington
Scientific Division
Cooper Biomedical, Inc.
One Technology Court
Malvern, PA 19355
(215) 251-2000

Costar
205 Broadway
Cambridge, MA 02139
(800) 492-1110

Difco Laboratories Ltd.
P.O. Box 14B
Central Avenue
East Moseley, Surrey, England

E. I. du Pont de Nemours & Co. (Inc.)
New England Nuclear Research Products
549 Albany Street
Boston, MA 02118
(800) 225-1572

Eastman Kodak Company
Biomedical Imaging
343 State Street
Rochester, NY 14650
(716) 724-1073

Eastman Kodak Company
Laboratory and Research Products Division
343 State Street
Rochester, NY 14650
(716) 724-1073

E-C Apparatus Corporation
3831 Tyrone Boulevard, North
St. Petersburg, FL 33709
(813) 344-1644
Telex 51-4736 HALA

Electronucleonics, Inc.
P.O. Box 451
Oak Ridge, TN 37830
(615) 483-1429

E-Y Laboratories, Inc.
107 North Amphlett Boulevard
San Mateo, CA 94401
(415) 342-3296

Falcon Labware
Becton Dickinson Labware Division
1950 Williams Drive
Oxnard, CA 93030
(800) 235-5953

Fisher Scientific
52 Fadem Road
Springfield, NJ 07081
(201) 379-1400

Flow Laboratories, Inc.
7655 Old Springhouse Road
McLean, VA 22102
(703) 893-5925

Forma Scientific/CCI
P.O. Box 649
Marietta, OH 45750
(614) 373-4763

Gibco/BRL
8400 Helgerman Court
Gaithersburg, MD 20877
(301) 258-8263

Gibco Laboratories
3175 Staley Road
Grand Island, NY 14072
(800) 828-6686
Telex 91-64077

519 Aldo Avenue
Santa Clara, CA 9505
(408) 988-7611
Telex 34-6381

Health Products, Inc.
Subsidiary of Pierce
P.O. Box 117
Rockford, IL 61105
(800) 336-1971

Hoefer Scientific Instruments
654 Minnesota Street
P.O. Box 77387
San Francisco, CA 94107
(415) 282-2307

Hyclone Laboratories
1725 South State Highway 89-91
Logan, UT 84321
(801) 753-4584

Ilford, Ltd.
Mobberley, Knutsford
Cheshire WA16 7HA, England
UK 565-50000

ISCO
4700 Superior Street
Lincoln, NE 68504
(402) 464-0231

Jackson Immunoresearch Laboratories, Inc.
P.O. Box 683
Route 41
Avondale, PA 19311
(215) 268-3026

LKB
260 North Broadway
Hicksville, NY 11801
(516) 433-0115
Telex 125160

4001 West Devan Avenue
Chicago, IL 60646
(312) 286-8143
Telex 25-5136

Mallinckrodt
Science Products Division
P.O. Box 5840
St. Louis, MO 63134

Medical Specialties, Inc.
P.O. Box 3485
Baltimore, MD 21225
(301) 467-2555

Miles Scientific, Division of Miles
Laboratories, Inc.
2000 North Aurora Road
Naperville, IL 60566
(312) 983-5700

Millipore Corporation
80 Ashby Road
Bedford, MA 01730
(617) 275-9200

Nalge Company
P.O. Box 365
Rochester, NY 14602
(716) 586-8800

Pharmacia Fine Chemicals AB
800 Centennial Avenue
Piscataway, NJ 08854
(800) 558-0005

Pierce Chemical Company
P.O. Box 117
Rockford, IL 61105
(815) 968-0747
Telex 91-631-3419

Polysciences
400 Valley Road
Warrington, PA 18976
(215) 343-6484

Protein Databases, Inc.
405 Oakwood Road
Huntington Station, NY 11746
(516) 673-3939

Rainin Instrument Company
Mack Road
Woburn, MA 01801
(617) 935-3050

Schleicher & Schuell, Inc.
10 Optical Avenue
Keene, NH 03431
(603) 352-3810

Sigma Chemical Co.
3500 DeKalb Street
St. Louis, MO 63118

Sigma Chemical Co.
P.O. Box 14508
St. Louis, MO 63178
(800) 325-3010 (ordering)
(800) 325-8070 (technical)
Telex 434475

Vector Laboratories
1429 Rollins Road
Burlingame, CA 94010
(415) 348-3737

Vickers Instruments, Inc.
P.O. Box 99
Malden, MA 02148
(617) 324-0350

The Virtis Company, Inc.
Route 208
Gardiner, NY 12525
(914) 255-5000

West Coast Scientific
1287 66th Street
Emeryville, CA 94608
(415) 654-2665

Zymed Labs, Inc.
Suite 5
52 South Linden Avenue
South San Francisco, CA 94080
(415) 871-4494

Index